P9-BXW-085

WITHDRAWN

USING THE
MATHEMATICAL
LITERATURE

BOOKS IN
LIBRARY AND INFORMATION SCIENCE

A Series of Monographs and Textbooks

EDITOR
ALLEN KENT

Director, Office of Communications Programs
University of Pittsburgh
Pittsburgh, Pennsylvania

Additional volumes in preparation

USING THE MATHEMATICAL LITERATURE

A PRACTICAL GUIDE

Barbara Kirsch Schaefer
Visiting Lecturer
School of Library and Information Science
State University College of Arts and Science
Geneseo, New York

MARCEL DEKKER, INC. New York and Basel

Lib. Sc
Ref
QA
41.7
S3

Library of Congress Cataloging in Publication Data

Schaefer, Barbara Kirsch, [Date] 1932 —
  Using the mathematical literature.

  (Books in library and information science ; v. 25)
  Includes index.
  1. Mathematical literature. I. Title. II. Series.
QA41.7.S3   510'.7'2   78-24537
ISBN 0-8247-6675-X

COPYRIGHT © 1979 by Marcel Dekker, Inc.   All Rights Reserved

Neither this book nor any part may be reproduced or transmitted in
any form or by any means, electronic or mechanical, including photo-
copying, microfilming, and recording, or by any information storage
and retrieval system, without permission in writing from the publisher.

Marcel Dekker, Inc.
270 Madison Avenue, New York, New York 10016

Current printing (last digit):
10 9 8 7 6 5 4 3 2 1

Printed in the United States of America

597462

TO PAUL

PREFACE

Surveys to determine the information needs and practices of mathematicians reveal the striking fact that many use the literature only infrequently. Whatever the reasons for this reluctance to use the literature, whether lack of knowledge about the existing literature, preference for other channels of communication, or simple disinclination, it is clear that many fail to appreciate the valuable services that are available.

The purpose of this book is to provide an insight into the vast and varied amount of mathematical literature and to act as a guide to its exploitation.

The emphasis of the book is on describing the different types of publications, citing selected titles as illustrations of each type. The book is not intended to be a comprehensive bibliography of the mathematical literature.

The book is addressed to three categories of users: (1) students, teachers, and practitioners of the mathematical sciences; (2) scientists in other fields who may use mathematics in their work; and, (3) students and practitioners of library science.

Unless otherwise noted, all the titles listed were in print at the beginning of 1977. Most of the literature discussed should be found in college libraries and large public libraries. It may be necessary to obtain some of the more esoteric journals and foreign language publications through interlibrary loan.

I wish to acknowledge the kind cooperation received from
Dr. Robert Bartle, Executive Editor of <u>Mathematical Reviews,</u>
Dr. Truman Botts, Executive Director of the Conference Board of the
Mathematical Sciences, and Dr. Gordon Walker, Executive Director
of the American Mathematical Society, who also granted permission
to reproduce material from <u>Mathematical Reviews</u>.

Geneseo, New York   Barbara K. Schaefer
March 1977

CONTENTS

USING THE MATHEMATICAL LITERATURE

Chapter 1

INTRODUCTION

Mathematics has been described as a collection of mathematicians doing mathematics. This circular statement typifies the difficulty of obtaining a satisfactory definition of the subject. Any two mathematicians, when asked to give a definition of mathematics, would disagree in significant ways, and a day later each would disagree with what he had said the day before [1].

Mathematicians have always found it impossible to talk intelligently to nonmathematicians about their work, and they find it increasingly difficult even to talk to one another. This is because the subject has undergone major internal transformations in recent years, the number of mathematicians has grown rapidly, and the pace and quantity of research has accelerated.

One characteristic of contemporary mathematics is that is has become more abstract. Mathematical research has become more preoccupied with mathematics per se and less concerned with other parts of science [2]. Paradoxically, research in pure mathematics is providing more and more of the methodological and conceptual tools required by modern science. The current penetration of mathematical methods into not only science and technology, but also the worlds of government, industry, and business, amounts to a virtual "mathematization of culture" [3].

It is ironic that while our society makes unprecedented demands on mathematics, little attention has been paid to the literature of this discipline, or to information systems in mathematics. This is in sharp contrast to the numerous studies and systems devoted to

other fields in science and technology. Perhaps a reason for this
is that although everyone uses mathematics, relatively few people do
mathematics [4]. The number of earned doctorates in mathematics provides a rough estimate of the number of people who do mathematics.
The number of Ph.D. degrees awarded in 1975/76 by U.S. and Canadian
universities was 1,046. (This figure includes degrees in all the
mathematical sciences: pure mathematics, statistics, computer science, operations research, other areas of applied mathematics, and
mathematics education.) This is not a large number compared to figures of over 4,000 in the physical sciences and about 3,000 in the
biological sciences. More significant is the growth rate within
mathematics. Between 1950 when 160 degrees were awarded and 1976 the
output of Ph.D.s increased at an annual rate of approximately eigh-
teen per cent [5].

With this increase in the number of research mathematicians,
there has been a corresponding increase in the mathematical literature. This has caused problems for the individual in keeping abreast
of his field and has taxed the existing apparatus for bibliographic
control of the literature. For example, the number of abstracts
in Mathematical Reviews, the principal abstracting journal of mathematics, has been doubling every eight or ten years [6]. Currently,
there is a time lag of up to twenty-four months before an article
or book is abstracted in Mathematical Reviews.

Various studies and surveys conducted by the mathematical community show that the information needs in the mathematical sciences
are met by a number of nearly independent activities of the professional societies, universities, and commercial publishers. Recent
attempts. to create a unified, comprehensive information system for
all of the mathematical sciences have been unsuccessful for a number
of reasons: (1) the various aspects of mathematics, which are discussed in Chapter 3; (2) the various information needs of mathematicians, which are discussed in Chapter 5; (3) satisfaction with
the status quo on the part of some; (4) lack of interest on the part
of others; and (5) financial considerations [7].

There is then, at present, no single information system, but
rather a collection of information services, which consist of "primary" and "secondary" publications. Primary publications constitute
original sources; that is, publications containing new material or
new presentations and discussions of known material. This category
includes the research journals, monograph series, proceedings of
symposia, and translation series. Secondary publications include
abstracting and reviewing journals, and indexes, which provide access to the primary publications; general works of reference, such
as encyclopedias, dictionaries, handbooks, mathematical tables, and
directories; and miscellaneous nonserial books, such as histories,
collected works of individual mathematicians, and other compilations.

In this guide these two categories of publications have been
regrouped into three divisions as follows: (1) journals; (2) books;
and (3) reference books. This reflects the order in which they are
used. The secondary access publications are discussed in connection with the primary publications to which they provide access:
abstracts and indexes to journals with the primary journals; bibliographies, indexes, and reviews of books with books. The remaining
secondary publications are discussed in the chapter on reference
books.

Two categories of material that are not included in this guide
are textbooks below the senior/graduate level, and mathematical
popularizations, games, and recreations. Textbooks are covered by
the "Telegraphic Reviews" section of The American Mathematical
Monthly, by the library reviewing magazine Choice, by the book review sections of many mathematical and scientific journals, and by
the Basic Library List of the Committee on the Undergraduate Program
in Mathematics of The Mathematical Association of America (1965).
Popular works can be located in general bibliographies, indexes, and
publishers' catalogs.

A recent publication should be mentioned here, however, because
it is unique to the subject of mathematics. The Annotated Bibliography of Expository Writing in the Mathematical Sciences, edited
by M. P. Gaffney and L. A. Steen, The Mathematical Association of

America, 1976, is a listing of over 1,100 articles and a few books
of general interest that appear in the widely scattered expository
literature of the mathematical sciences. The entries in this bibliography are arranged by subject, supplemented by an author index.
Within each subject area, entries are grouped according to level
(general, elementary, advanced, and research) and within each level
they are arranged alphabetically. Each article is listed with full
bibliographic data and annotation. The primary purpose of this
bibliography is to provide reading lists to be used in undergraduate
mathematics courses, but the material that is included and the
grouping of entries according to levels of mathematical sophistication makes it useful to the mathematically oriented layman, the
scientist, and the mathematician, as well as the student.

The major portion of this guide, Chapters 7 to 12, is devoted
to the current literature of pure, or core, mathematics. The literature of the applied mathematical sciences presents some unique problems and is, therefore, discussed separately in Chapters 13 to 15.

The following five chapters are intended to serve as background
for the discussion of the different categories of publications that
begins with Chapter 7. Chapter 2 is a brief historical survey of
the literature up to the present time. Chapters 3 to 5 deal with
the nature of contemporary mathematics, and the nature and use of
the literature. Chapter 6 on library classification and cataloging
may be of use to the reader who is unfamiliar with library operations.

REFERENCES

1. Alfred Adler, "Mathematics and Creativity," The New Yorker,
    February 19, 1972, pp. 39-45.

2. Marshall Stone, "The Revolution in Mathematics," The American
    Mathematical Monthly, 67, 715-734, (1961).

3. Committee on Support of Research in the Mathematical Sciences,
    Report of the Committee, The Mathematical Sciences, National
    Academy of Sciences, Washington, D. C., 1968.

4. J. D. Mack, "A Mathematical Problem," Wilson Library Bulletin,
    42, 610-612, (1968).

5. W. H. Fleming, "Report on 1976 Survey of New Doctorates," Notices
    of the American Mathematical Society, 23, 318-341, (1976).

6. R. L. Wilder, "Trends and Social Implications of Research," Bul-
    letin of the American Mathematical Society, 75, 891-906, (1969).

7. Committee on a National Information System in the Mathematical
    Sciences, Information Needs in the Mathematical Sciences, Con-
    ference Board of the Mathematical Sciences, Washington, D. C.,
    1972.

Chapter 2

A BRIEF HISTORY OF THE MATHEMATICAL LITERATURE

Throughout the long history of mathematics, the records and communications of mathematicians have taken a variety of forms, depending not only upon the state of the art, but also upon the information needs and the technology at a given time. The investigation of the literature of mathematics may be placed in perspective by relating the development of the literature to the development of mathematics itself.

For convenience in discussing the subject, mathematics history is divided into nine chronological periods. These historical divisions are only approximations, of course, since ideas tend to overlap and intermingle with the passage of time. They are intended simply as an aid in analyzing the characteristics and accomplishments of mathematics, and the literature of mathematics, at various stages in history.

I. PREHISTORY

Knowledge about mathematics in prehistoric times is based largely on conjecture, for the beginnings of the subject are older than the art of writing. It is generally believed, however, that during that period two general concepts began to emerge: quantity, and form. The concept of quantity is thought to have started from attempts to compare collections of objects by counting, and slowly evolved into a variety of primitive number systems. These number systems became more refined as exchange and barter became important

7

and eventually developed into arithmetic. The concept of form probably began as primitive art; as decoration for pottery, clothing,
and buildings. It is thought that the mathematical aspects of form
(that is, geometry) did not emerge until much later, as a practical
aid in mensuration [1].

II.   5000 B.C. TO 600 B.C.

By the time of the historic period, general methods of calcu-
lation and measurement had been developed to meet practical needs.
Writing had also been developed by this time, enabling early societies to record these methods for future use.

More is known of the mathematics of the early civilizations of
the Near East than of other areas, since the Babylonians wrote on
baked clay tablets, which were almost indestructible, and the Egyptians used papyrus, which survived fairly well in the dry climate
of northern Africa. Records from this period indicate that by 4241
B.C. the Egyptians had developed a calendar, the Sumerians had a
mercantile arithmetic by 3000 B.C. and a well-developed number system by 2100 B.C., and by 1950 B.C. the Babylonians had established
an elementary algebra. The Babylonians also had a geometry, consisting of formulas for simple areas and volumes, and including the
rule for triangles that is now called the Pythagorean Theorem.

Most of these facts have been inferred from ceremonial and
astronomical writings, and from inscriptions on tombs and monuments.
The earliest source of information of a purely mathematical nature
is the Ahmes Papyrus, written in 1650 B.C. Named for the scribe
Ahmes, who stated that he was copying an earlier work that had been
written about 1800 B.C., it is a practical handbook of the mathematical knowledge of that time. It is also known as the Rhind
Papyrus, after A. Henry Rhind, the archaeologist who brought the
papyrus to England in the nineteenth century. A two-volume edition
of the Rhind Mathematical Papyrus was published under the auspices
of The Mathematical Association of America in 1951. Volume 1 contains a free translation, commentary, and bibliography of Egyptian

mathematics. Volume II contains photographic plates, text, and introduction, and literal translation. It is now, unfortunately, out of print, but can probably be found in most university libraries.

Relatively little is known of Chinese and Indian mathematics of this period, because these peoples wrote on bark or bamboo; materials that were far more perishable than clay tablets or papyrus. The burning of all existing books and the burying alive of all protesting scholars by order of the Emperor Shi Huang-ti of the Ch'in Dynasty in 213 B.C. is another reason why there are so few records from this period in Chinese history. Transcriptions of some ancient treatises have survived, however, including the Chou-pei and Chiu ch'ang Suan-shu (Nine Chapters on the Mathematical Arts). Both books date from the period of the Han Dynasty (206 B.C. to A.D. 220), but some of the material in them dates from an earlier period. The Chou-pei deals with astronomy and mathematics, and is of interest because it contains a discussion of the Pythagorean Theorem. The Nine Chapters deals exclusively with mathematics, consisting mainly of a set of problems with general rules for their solution.

Even less is known of Indian mathematics of this period. The òldest extant texts date from the first centuries A.D. It can only be said that there is evidence that the ancient Hindus had a workable number system and an elementary geometry [2].

III. 600 B.C. TO A.D. 400

Mathematics of the pre-Hellenic period was merely a tool in the form of disconnected, simple rules derived from experience and immediately applicable to daily life. Conclusions were established empirically and deductive reasoning was completely lacking. Mathematics did not develop as an organized, independent, and reasoned discipline until the blossoming of Greek civilization [3].

At the beginning of this period in history men such as Thales of Miletus (ca. 640 to ca. 546 B.C.), Pythagoras (ca. 570 to ca. 500 B.C.), and Zeno of Elea (ca. 450 B.C.) began, for the first time, to speculate about the rules of mathematics. The subsequent develop-

ment of logical principles and axiomatic methods by Plato (429 to
348 B.C.) and Aristotle (384 to 322 B.C.) and their followers contributed to the emergence of mathematics as a science studied for
its own sake.

Probably the most famous figure of this period was Euclid (ca.
300 B.C.), whose greatest contribution was the thirteen-volume
treatise, the Elements. The Elements incorporated all the essential
accumulated mathematical knowledge of the time, organized into a
system logically deduced from a single axiomatic foundation. The
works of Euclid were so comprehensive that they superseded all previous writings, and it is probably for this reason that very few
pre-Euclidean Greek manuscripts were preserved [4].

During the Alexandrian period, from about 300 B.C. to A.D. 400,
Alexandria became the mathematical center of the Western world. Two
important figures connected with the famous School of Alexandria
were Archimedes of Syracuse (287 to 212 B.C.) and Apollonius of
Perga (ca. 260 to ca. 210 B.C.). Archimedes, an astronomer, physi-
cist, and applied and speculative mathematician, made major contributions to number theory, algebra, and geometry. His treatise, On
the Equilibrium of Planes, is one of the oldest extant books on
physical science. Of the many treatises by Apollonius, called "the
great geometer" because of his important contributions to synthetic
geometry, only one work has survived: the Conics [1].

After Apollonius, Greek mathematics began its decline, along
with the rest of Greek civilization. For the next 600 years, only
two important writers stand out among the minor ones. The first,
Ptolemy (Claudius Ptolemaeus, ca. A.D. 85 to ca. 165), wrote a comprehensive treatise on astronomy, the Almagest, which extended computational mathematics to plane trigonometry and the beginnings of
spherical trigonometry. The second man, Diophantus of Alexandria
(ca. A.D. 275), wrote the Arithmetica, an important contribution
to the development of number theory; six of these books have survived [5].

IV. A.D. 400 TO 1400

With the end of Greek civilization, the center of mathematical activity shifted to India, Central Asia, and the Arabic countries.

The Hindus achieved significant results in algebra and arithmetic, but their most important contribution was the development of the numeration system we use today: a place system based on ten and including a symbol for zero.

Perhaps the greatest contribution of the Moslems was preserving the continuity of mathematical thought. They translated most of the important Greek and Hindu manuscripts into Arabic and spread them in their travels throughout the Arab world. However, a Moslem scholar of this period who deserves special mention is Mohammed ibn Musa al-Khowârizmî (ca. 825). He wrote two books, one on arithmetic and the other on algebra, which were translated into Latin in the twelfth century. The title of the first in translation is Algorithmi de numero Indorum (literally "Al-Khowârizmî on Indian numbers"), which is the source of the term "algorithm." The second book is entitled Al-jebr al-muqabala (literally "Restoration and Opposition"), and by latinization the key word in the title became "algebra" [6].

There was virtually no mathematical progress in Europe during this time. Throughout the period of the Roman Empire and until the fifteenth century, the language of instruction in European schools was Latin. Despite their contributions in many other areas, the Romans did not produce any significant mathematics. The only noteworthy Roman mathematician was Anicius Manlius Severinus Boetius (ca. 480 to 524), whose works included an arithmetic, Institutis arithmetica, a geometry, Geometry, and a treatise on music (considered to be a part of mathematics at that time). His texts were used by the European monastic schools until the twelfth century.

As trade and travel expanded, the Europeans came into contact with the Arabs of the Mediterranean area and the Near East and with the Byzantines of the Eastern Roman Empire, from whom they learned about the Greek mathematical classics. During the twelfth century

the Greek works that had been translated into Arabic a few centuries
before were translated into Latin.

The next few hundred years were a time of absorption of the
Greek and Arabic mathematics. Mathematical creativity in the West
did not resume until the fifteenth century. The emerging European
civilization, which had seemed so promising with the founding of
the first universities in the thirteenth century, was postponed by
the Hundred Years' War (1337 to 1453) and the Black Death (1347 to
1351) [2].

V. THE FIFTEENTH AND SIXTEENTH CENTURIES

In response to the increasing demands of astronomy, navigation,
trade, and surveying, mathematics of this period was concerned prin-
cipally with computation. The leading mathematician of the fifteenth
century, Johannes Müller (1436-1476), was also an astronomer and he
wrote De triangulis omnimodis, the first treatise to be devoted
solely to trigonometry. The foremost scientist of this age, Leonardo
da Vinci (1452-1519), was concerned with the application of geometry
to the physical sciences and also to art. His writings on perspective are contained in his Trattato della pittura, compiled in 1651
by some unknown author.

There also occurred during this period two events that laid the
groundwork for future mathematical developments: the invention of
movable-type printing and the beginning of writing in the vernacular.

The first printed books on mathematics were a commercial arithmetic (1478) and a Latin edition of Euclid's Elements (1482). Of
greater mathematical significance was the publication in 1494 of
Summa de Arithmetica, by Luca Pacioli (1445 to ca. 1514). This
book, written in Italian, was a compilation of all the arithmetic,
algebra, and trigonometry known at that time.

Robert Recorde (ca. 1510-1558), called "the father of English
mathematics," published four books on mathematics, written in English: The Castle of Knowledge, an exposition on the Copernican theory in astronomy; The Ground of Artes, an arithmetic; The Pathwaie

to Knowledge, an abridgement of Euclid's Elements; and The Whetstone
of Witte, an algebra [5].

VI. THE SEVENTEENTH CENTURY

As part of the scientific explosion of the seventeenth century,
mathematics experienced a period of unprecedented growth. The practical demands of society, the spread of education, and the general
intellectual climate of the time contributed to this growth. The
number of people engaged in mathematical activity became so large
that only a few of the outstanding mathematicians and their major
works can be mentioned.

John Napier (1550-1617), a Scottish mathematician, published
Mirifici Logarithmorum Canonis Descriptio (1614), in which he developed the theory of logarithms.

René Descartes (1596-1650), the French philosopher-scientist,
sought to unify all of science by means of logic and mathematics.
One of the appendixes to his famous Discourse on Method (1637),
entitled "La Geometrie," was the first publication of analytic geometry (the application of algebraic methods to geometry).

Much of the work of Gérard Desargues (1593-1662), considered
the founder of modern projective geometry, was eclipsed during his
lifetime by Descartes' writings. Probably his best-known work is
Brouillon project d'une atteinte aux événemens des rencontres du
cône avec un plan (1639).

Pierre de Fermat (1601-1665), who is probably best-known for
his work in number theory, also made important contributions to
analytic geometry, the theory of probability, and the technical
foundations of calculus, prior to Newton. Fermat published only a
few papers; most of his results are known through letters he wrote
to friends and from notes he made in the margins of books.

Another Frenchman, Blaise Pascal (1623-1662), did work in geometry (Essay on Conics, 1640), and also was the cocreator with Fermat
of probability theory [2].

One of the most significant contributions of the seventeenth
century was the invention of differential and integral calculus,
which led to the development of analysis, one of the major branches
of mathematics. Calculus arose as a method for solving problems in
the new science of mechanics and old problems in geometry. It may
be characterized as the gateway between elementary and advanced
mathematics; a knowledge of calculus is essential to any real understanding of physics and related branches of technology and a pre-
requisite to the study of analysis. As is the case so often in mathematical creativity, calculus was developed independently in England
by Isaac Newton (1642-1727) and in Germany by Gottfried Wilhelm
Leibniz (1646-1716). This fact engendered one of the most bitter
partisan feuds in mathematical history, with admirers of each man
hurling charges of plagarism at the other.

Newton's first publication involving his calculus was his famous Philosophiae Naturalis Principia Mathematica, 1687, (The Mathematical Principles of Natural Philosophy). His later monographs on
the subject are: De Analysi per Aequationes Numero Terminorum In-
finitas, 1669, (An Analysis by Means of Equations with an Indefinite
Number of Terms); Methodus Fluxionum et Serierum Infinitarum, 1671,
(Method of Fluxions and Infinite Series); and Tractatus de Quadra-
tura Curvarum, 1676, (Quadrature of Curves).

Leibniz's results, as well as the development of his ideas, are
contained in hundreds of pages of notes made from 1673 on, but never
published by him. In 1684 he began publishing a few papers on calculus in the journal Acta Eruditorum, and in 1714 he wrote Historia
et Origo Calculi Differentialis, in which he gave an account of the
development of his own thinking on the subject [4].

In addition to the expansion of mathematical content and activity, the seventeenth century is significant for developments that
occurred with respect to communication among mathematicians. Prior
to the late sixteenth century, mathematics was created by individuals
and small groups headed by one or two prominent leaders. The results
were communicated orally, through personal correspondence, and occa-
sionally written up in texts; first in manuscript form and later as

printed books. As more people began to participate in mathematical
research, the desire for exchange of information and for the stimu-
lus of meeting others with the same intellectual interests resulted
in the founding of scientific societies. These societies were important not only in making possible direct contact and exchange of
ideas, but also because they eventually supported journals, which
by the end of the eighteenth century became the accepted medium for
publication of new research results. The oldest mathematical society
still in existence is the Mathematische Gesellschaft in Hamburg,
Germany, founded in 1690 as the Kunstrechnungsliebende Societät,
and continued from 1790 to 1876 as the Gesellschaft zur Verbreitung
der Mathematischen Wissenschaften. Its publications were entitled
Jahresbriefe or Jahres-Berichte or Berichte from 1723 to 1878 and
its current journal, which began in 1873, is entitled Mitteilungen
[7].

Another seventeenth-century development was the establishment of
Royal Academies, subsidized by the rulers of the time. These academies sponsored research and also supported journals, in which they
published the results of their research.

VII. THE EIGHTEENTH CENTURY

Development of the new theory of calculus played an important
role in mathematical research of the eighteenth century. In connection with problems of mechanics, physics, and technology, many other
new fields emerged; such as, infinite series, the theory of differential equations, calculus of variations, and the application of
analysis to geometry, which later became differential geometry.

The centers for research during this century were the Royal
Academies. Prior to 1800, universities played only a minor role in
research and the great mathematicians were attached to the academies.
The most prominent academies were at Berlin, London, Paris and
St. Petersburg.

Leonhard Euler (1707-1783), who was associated with both the
Berlin Academy and the St. Petersburg Academy during his career, was

the most productive mathematician of the century. He wrote about
nine hundred books and papers, including works on mechanics, music,
and astronomy, as well as mathematics. Euler is considered the
creator of analytic mechanics (as opposed to the older geometrical
mechanics), based on his achievements in establishing calculus as a
purely analytical theory. His Mechanica (1736) is a major contribution to the analytical foundation of mechanics. Euler's texts
on mechanics, algebra, mathematical analysis, analytic and differential geometry, and calculus of variations were standard works for
a hundred years or more. Three landmark texts in calculus are:
Introductio in Analysin Infinitorum (1748), the first connected presentation of calculus and elementary analysis; Institutiones Calculi
Differentialis (1755); and, the three-volume Institutiones Calculi
Integralis (1768-1770) [1].

French mathematicians continued to be very active during the
eighteenth century and some of the important names and works from
that country are: Traité de dynamique (1743) and Reflexions sur la
cause générale des vents (1747), by Jean d'Alembert (1717-1783);
Mecanique analytique (1788) and Theorie des fonctions analytiques
(1797), by Joseph Louis Lagrange (1736-1813); Eléments de géométrie
(1794) and Essai de la théorie des nombres (1798), by Adrien-Marie
Legendre (1752-1833); and, Feuilles d'analyse appliquée à la géométrie (1795), by Gaspard Monge (1746-1818). An important work,
published between 1799 and 1825, is the five-volume Mécanique céleste,
by Pierre Simon Laplace (1749-1827). This work reviewed, unified,
and greatly extended all previous work of Newton, d'Alembert, Euler,
Lagrange, and Laplace, himself, in the fields of probability and
celestial mechanics [1], [2].

VIII. THE NINETEENTH CENTURY

The modern era in mathematics is considered to have begun with
Carl Friedrich Gauss (1777-1855), the German mathematician who dom-
inated nineteenth century mathematics. He contributed to almost
every branch of mathematics, but is perhaps best known for his work

in the theory of numbers, represented by his <u>Disquisitiones Arithme-</u>
<u>ticae</u> (1801).

During the 1800s mathematical activity continued at an ever-
increasing rate. Every major branch of mathematics underwent pro-
found changes in terms of expansion of subject matter and the open-
ing of new fields, as well as the extension of older ones. Conse-
quently, mathematicians were forced to specialize in one branch of
the subject. It has been said that Gauss was one of the last men
to know mathematics as a whole [5].

Algebra, which previously was little more than generalized
arithmetic, became more abstract, with wider applications, during
this period. Early work in this area was done by the Irish mathematician William Rowan Hamilton (1805-1865) and the German mathematician Hermann Grassmann (1809-1877). Important works of Hamilton
are <u>Lectures on Quaternions</u> (1853) and <u>Elements of Quaternions</u>
(1866), and of Grassmann, <u>Die Lineale Ausdehnungslehre</u> (1844). Two
important English algebraists were Arthur Cayley (1821-1895), the
originator of matrix theory, and James Joseph Sylvester (1814-1897).
Cayley's writings are found in his <u>Collected Mathematical Papers</u>, 13
volumes, Cambridge University Press, 1889-1897. Sylvester, who
taught at Johns Hopkins University from 1876 until his return to
England in 1884, published in various journals. He is also important for having initiated research in pure mathematics in·the United
States and for founding the first American mathematics journal, the
<u>American Journal of Mathematics</u>. Also in the field of algebra, work
was done in the theory of groups, notably by Felix Klein (1849-1925)
in Germany, and by Marius Sophus Lie (1842-1899) in Norway. Both
men published widely in journals and Klein also wrote books on the
theory, history, and teaching of mathematics. Two of his books are
<u>Famous Problems of Elementary Geometry</u> (1895; English translation
1930), and <u>Elementary Mathematics from an Advanced Standpoint</u> (2
volumes, 1907-1908; English translation 1932-1940). Lie's work in
continuous transformation groups was codified in a number of books,
edited with the aid of his pupils Georg Scheffers (1866-1945) and
Friedrich Engel (1861-1941): <u>Theorie der Transformationsgruppen</u>,

1888-1893; Differentialgleichungen, 1891; Kon   iche Gruppen,
1893; and, Berührungstransformation, 1896. A.   .r man working in
algebra at this time was the French mathematician Évariste Galois
(1811-1832). One of the more romantic figures in mathematics history, Galois was killed in a duel in 1832, short' before his twenty-
first birthday. His unpublished notes, that were later deciphered
and revised, turned out to be the theory of groups, the foundations
of modern algebra, and modern geometry. Most of Galois' papers were
later published in the Journal de mathematiques in 1846 [2].

In the field of analysis, three men are acknowledged to be the
founders of the theory of complex functions: Augustin Louis Cauchy
(France, 1789-1857), Bernhard Riemann (Germany, 1826-1866), and Karl
Weierstrass (Germany, 1815-1897). Cauchy was a prolific writer,
whose over seven hundred papers covered all branches of mathematics.
His first significant paper in complex function theory was "Mémoire
sur la théorie des intégrales définies," read to the Paris Academy
in 1814. The paper considered his most important was "Mémoire sur
les intégrales définies prises entre des limites imaginaires,"
written in 1825 but not published until 1874. Riemann's doctoral
thesis of 1851, "Grundlagen für eine allgemeine Theorie der Func-
tionen einer veränderlichen complexen Grösse," is a basic paper in
complex function theory. Weierstrass did not publish his results
at the time he first achieved them. His research became known only
in the late 1890s when he published his collected Werke. Also in
analysis, the theory of elliptic functions was developed by Niels
Henrik Abel (Norway, 1802-1829) and Carl Gustav Jacob Jacobi (Germany,
1804-1851). The paper that contains Abel's famous theorem on the
sum of integrals of algebraic functions was first submitted to the
Academy of Sciences in Paris in 1826 for publication in its journal.
The manuscript was either disregarded or mislaid, however, and it
was not published until 1841, after Abel's death. A key work in
elliptic functions is Jacobi's book Fundamenta Nova Theoriae Func-
tionum Ellipticarum (1829) [3].

At the beginning of the nineteenth century, a period of criticism, systematization, and laying of foundations occurred in analy-

sis. Through the efforts of such people as Gauss, Abel, Cauchy, Riemann, and Weierstrass, mentioned above, and later Leopold Kronecker (1823-1891), Richard Dedekind (1831-1916), and Georg Cantor (1845-1918), analysis was placed upon a firmer basis than had been provided by the first creators of its powerful methods. In particular, exact and general definitions were given for the basic concepts of real number, variable, function, limit, and continuity. The establishment of a firm logical foundation did not represent the end of the development of analysis; on the contrary, it provided impetus for further development. It was in connection with his investigations in analysis that Cantor developed a new mathematical theory, the theory of infinite sets, whose ideas and methods have influenced not only analysis but all branches of mathematics. During the twentieth century, analysis took new directions, resulting in the theory of functions of a real variable, theory of approximation of functions, qualitative theory of differential equations, theory of integral equations, and functional analysis. In general, these newer branches of analysis are called modern analysis in contradistinction to the earlier so-called classical analysis [6].

The introduction of analytic geometry by Fermat and Descartes in the seventeenth century constituted the first development of that branch of mathematics since the Elements of Euclid (third century B.C.). Nineteenth century developments in geometry affected not only that field, but all the branches of mathematics. Early in the century, three men in different countries all published consistent non-Euclidean systems of geometry, each independently of the other two. In Russia, Nicolai Lobachevsky (1793-1856) published a series of journal articles between 1829 and 1837, and in 1840 he published a book in German: Geometrische Untersuchungen zur Theorie. His Pangéométrie, published in 1855, was a completely new exposition of his geometry. In Hungary, János Bolyai (1802-1860) wrote a paper entitled "The Science of Absolute Spaces," which was published in 1832. And in Germany, Bernhard Riemann, in 1886, published "Über die Hypothesen, welche der Geometrie zu Grunde Liegen' (On the Hypotheses Which Lie at the Foundation of Geometry). These men showed that the

axioms of Euclid were not self-evident truths, as had been assumed
for centuries, and that consistent geometries built on other axioms
were possible. This idea produced fundamental changes in notions
of physical and mathematical space, and the axiomatic method that
pervades all modern mathematics is one of the consequences of this
discovery [6].

Attempts to unify mathematics and logic have challenged mathematicians and philosophers throughout history. During the second
half of the nineteenth century there was a renewal of interest in
logic and foundations of mathematics, first stimulated by the work
of George Boole (1815-1864) and later by Georg Cantor's controversial
theory of sets (Mengenlehre). The Mathematical Analysis of Logic
(1847) and The Laws of Thought (1854), both by Boole, were the first
important mathematical studies of logic. They laid the foundation
for the subsequent development of mathematical logic (symbolic logic).
Cantor's theory is spread over numerous papers that appeared in
Mathematische Annalen and the Journal für Mathematik from 1874 to
1897 [8].

As mathematics expanded and changed during the nineteenth century, so too did the agencies for the propagation of results. First,
the universities participated in research, the writing of textbooks,
and the systematic training of mathematicians. The dominance of the
royally supported academies began to decline at this time and research
became an important function of the universities [3].

Second, there was a vast increase in the number of journals.
The first purely mathematical journal, Annales de Mathematiques Pures
et Appliquees, began in France in the nineteenth century. It was
published from 1810 until 1831. In Germany, Crelle's Journal began
in 1826 and is still being published. Among the numerous other titles
that started at this time are the Mathematische Annalen (Germany,
1868 to date), the Acta Mathematica (Sweden, 1882 to date), and the
American Journal of Mathematics (United States, 1878 to date).

Finally, another type of agency that has promoted mathematical
activity since the mid-nineteenth century is the national professional society. These societies hold regular meetings at which

papers are presented and each sponsors one or more journals. (See
Chapter 7 for a discussion of society journals.)

Occasional international meetings of mathematicians took place
during the nineteenth century, but the first of the present series
of international congresses of mathematicians was held in 1897. The
congress has met regularly every four years since 1900, except for
interruptions caused by the two World Wars. The Proceedings of each
congress have been published, usually by the mathematical society of
the sponsoring country [7].

IX. THE TWENTIETH CENTURY

Interest in the logical foundations of mathematics began to
spread rapidly, as a result of Boole's work in logic and Cantor's
theory of sets, a theory upon which it was believed all mathematics
could be based. In Principia Mathematica, a monumental two-volume
work that appeared during the years 1910-1913, the philosopher-
mathematicians Bertrand Russell (1872-1970) and Alfred North White-
head (1861-1947) attempted to express all of mathematics in a universal logical symbolism. David Hilbert (1862-1943), whose Grundlagen der Geometrie (Foundations of Geometry) is a complete revision
of Euclid's Elements using modern axiomatic methods, sought to unify
mathematics by finding a single provably consistent set of axioms
upon which all mathematics could be based. This goal was proved un-
attainable in 1931 when Kurt Gödel (1906-1978) published "Uber formal
unentscheidbare Sätze der Principia Mathematica und verwandter Sys-
teme I" (On Formally Undecidable Propositions of Principia Mathematica and Related Systems; English translation 1965). Another important work of Gödel's is The Consistency of the Axiom of Choice and
of the Generalized Continuum Hypothesis with the Axioms of Set Theory
(1940; revised edition 1951). Gödel demonstrated that in any axiom
system broad enough to contain all the formulas of a formalized
elementary number theory, there exist theorems (formulas) that can
neither be proved nor disproved within the system [9].

These difficulties in the foundations of mathematics have not yet been wholly resolved and mathematical logic is currently an active research field. At the same time, set theory, which has continued to develop as a separate field of study, has influenced all mathematics. The dominance of the set-theoretical point of view is one of the characteristics of contemporary mathematics and virtually all the concepts of contemporary mathematics are phrased in the language of set theory [6].

While mathematical logic and set theory are highly abstract subjects, closely related to philosophy, it is interesting to note that they have contributed to America's fastest-growing industry, the computer industry. In attempting to reduce mathematical reasoning to mere manipulation of symbols, using only stated axioms and rules of inference, early researchers in mathematical logic prepared for the development of digital computers. This is another example of what the history of science has shown time and time again: that it is impossible to predict what mathematical theories will turn out to be useful outside of pure mathematics [10].

The desire to consolidate and simplify previous results can be seen throughout mathematical history; Euclid's Elements is an example of this tendency. In the twentieth century, another attempt was begun in the 1930s to integrate all contemporary mathematics within a single framework by a group of French mathematicians writing under the name of "Bourbaki." Nicolas Bourbaki is the collective pseudonym used by an informal corporation of French mathematicians, numbering from ten to twenty men at any one time. This anonymous society is writing a comprehensive treatise on mathematics, starting with the most general basic principles and to conclude, presumably, with the most specialized application. The treatise, whose general title is Élémentes de Mathématiqué, is a survey of all mathematics from a sophisticated point of view. The Bourbaki presentation of each subject is systematic and thorough, often including a historical review of the subject and a set of exercises. The Bourbaki gadgetry includes inserts constituting a set of directions on the proper use of the treatise, foldout sheets that summarize important definitions and

assumptions, and a dictionary for each book that serves also as an
index to "Bourbachique" terminology. The project, published by
Hermann of Paris, got under way in 1939 and over thirty titles of
this monumental work have appeared to date. The number of volumes
is somewhat larger, since many of the earlier titles have been revised and issued as second and even third editions. The influence
of this series is equalled only by the amount of apocryphal stories
about Bourbaki, most of them perpetrated by the members of the group
[11].

At the International Congress of Mathematicians held in Paris
in 1900, David Hilbert lectured on mathematics in the new century.
He outlined twenty-three unsolved problems as a challenge for the
1900s. Not all of these problems have yet been solved, but attempts
to settle each one have led to important new results. As keen as
Hilbert's insight proved to be, even he could not predict at the
beginning of the century the developments that would occur in the
coming years.

Since the 1940s, methodological and conceptual tools of mathematics have been utilized at an ever-increasing rate by the physical
and social sciences. Within mathematics, itself, there has been an
increasing tendency to apply the methodology of one branch of mathematics to other branches. This has resulted in the continued development of existing fields and the creation of a number of new fields.
As a consequence, mathematics has been growing at an astonishing
rate. It has been estimated that of all the mathematicians who ever
lived, over 80 percent are alive today, and that more original mathematics has been produced since the end of World War II than in all
previous history. These developments have led many mathematicians
to regard the second half of the twentieth century as the golden
age of mathematics [12].

REFERENCES

1. C. B. Boyer, A History of Mathematics, Wiley, New York, 1968.

2. D. J. Struik, A Concise History of Mathematics, 3rd revised
    edition, Dover, New York, 1967.

3. Morris Kline, Mathematical Thought from Ancient to Modern Times,
    Oxford University Press, New York, 1972.

4. D. E. Smith, History of Mathematics, Volumes 1 and 2, Ginn,
    Boston, 1923.

5. W. P. Berlinghoff, Mathematics: The Art of Reason, Heath,
    Boston, 1968.

6. A. D. Aleksandrov, "A General View of Mathematics," in Mathe-
    matics: Its Content, Methods, and Meaning, edited by A. D.
    Aleksandrov, A. N. Kolmogorov, and M. S. Lavrent'ev, translated
    by S. H. Gould and T. Bartha, Volume I, M.I.T. Press, Cambridge,
    Mass., 1963, pp. 1-64.

7. "Mathematics, Societies of," in Encyclopaedia Britannica,
    Volume 14, 1973.

8. N. I. Styazhkin, History of Mathematical Logic from Leibniz to
    Peano, M.I.T. Press, Cambridge, Mass., 1969.

9. R. L. Wilder, Introduction to the Foundations of Mathematics,
    Wiley, New York, 1952.

10. Committee on Support of Research in the Mathematical Sciences,
    Report of the Committee, The Mathematical Sciences, National
    Academy of Sciences, Washington, D. C., 1968.

11. P. R. Halmos, "Nicolas Bourbaki," in Mathematics in the Modern
    World, Readings from Scientific American, Freeman, San Francisco,
    1968, pp. 77-81.

12. T. L. Saaty, The Spirit and Uses of the Mathematical Sciences,
    McGraw-Hill, New York, 1969.

Chapter 3

THE NATURE OF MATHEMATICS

Because of the long and essential connection between mathematics
and physics, mathematics is often classed with the physical sciences.
Until very recently its impact on society was almost exclusively
through its applications to astronomy, chemistry, physics, and engineering. But mathematics has features that distinguish it from all
experimental sciences; in fact, many mathematicians consider it to
be an art rather than a science.

The basic difference between the mathematical sciences and the
experimental sciences, including even the most theoretical branches,
lies in the extent to which mathematical research originates within
the body of mathematics itself. Physics, chemistry, etc.; are more
or less concerned with observable phenomena; mathematics deals rather
with the relationships between ideas. Mathematicians often receive
ideas from the outside world, sometimes through other sciences, but
once a mathematical problem or concept has been formulated, it takes
on a life of its own. In a simplified way, mathematics consists of
abstractions of reality, abstractions of abstractions of reality,
and so on. The amazing thing is that so often mathematics created
for its own sake turns out to be important for applications. For
example, Cayley believed that matrices, which he invented, would
never be applied to anything useful (and was happy about it). They
are now an everyday working tool of engineers, physicists, econo-
mists, and statisticians.

Another feature of mathematics is that in this century the
boundaries between the traditional branches (algebra, geometry,

analysis, etc.) have become less precise than formerly. Some of the
most important developments in contemporary mathematics are occurring
at the interfaces between different fields, leading to the creation
of entirely new fields. This has been a result of the recognition
of fundamental interrelationships among the branches of mathematics
and the application of the methodology of one branch to other branches,
mentioned in the previous chapter [1].

Research in mathematics, unlike many of the other sciences, is
typically done by individuals rather than by groups. While it often
happens that important discoveries are made by several people working
independently in different locations and that personal contact among
mathematicians in the same field is often an essential condition for
successful work, the actual work is almost always done by one person
thinking about a problem. Very few mathematical papers have more
than two authors.

Looking more closely at the characteristics that distinguish
the mathematical sciences, three aspects of mathematics can be identified: research, applications, and exposition. These aspects have
direct bearing on the information needs of mathematicians and on the
types of mathematical literature that are produced.

Mathematical research is in essence concerned with the discovery
and proof of theorems, and the logical connectives between theorems.
The relationship of research papers to the total body of literature
is to other articles in the same general field, and often a narrow
subfield. Thus, the information needs of a research mathematician
can be satisfied by access to a relatively small area of mathematics.
He can keep abreast of developments in his field fairly well by
reading the journals in his area of specialization, by using the
abstracts that cover the research literature, and by contact with
other people doing research in his field.

Applications of mathematics involve, on the other hand, important connections with the fields of application. Furthermore, these
applications may use techniques and research results from several
subfields of the mathematical sciences. The results of work in
applied areas may be either research theorems as in pure mathematics,

or they may report the successful solution of specific applied problems through the new use of known mathematical methods. Thus, the
literature dealing with applications is widely dispersed, often in
publications of the field of application, rather than in purely
mathematical publications. The applied mathematician is forced to
consult a greater number of journals and abstracts to locate needed
information.

The third facet of the mathematical sciences, exposition, involves bringing together the research results, both theorems and
applications, with special attention to consolidation of related
results, simplification, and development of relationships with a
general body of theory. This has been a somewhat neglected area of
mathematics and there is not a large body of expository literature.
This is due in part to the inherent difficulty of such a task, to
the problem of presenting mathematics to the nonspecialist, and to
the reluctance of many mathematicians to take time away from their
research [2].

It should be pointed out that in mathematics the term "exposition" is not synonymous with "popularization." Mathematics cannot
be popularized in the sense that it can be written about; the reader
must always "work out" the mathematics as he goes along. Nor does
"nonspecialist" mean "layman;" the term is used to describe a mathematician who is reading in an area that is not his area of specialty.
Whenever these terms are used in this book it is with these meanings.

REFERENCES

1. Committee on Support of Research in the Mathematical Sciences,
    Report of the Committee, The Mathematical Sciences, National
    Academy of Sciences, Washington, D.C., 1968.

2. Committee on a National Information System in the Mathematical
    Sciences, Information Needs in the Mathematical Sciences, Con-
    ference Board of the Mathematical Sciences, Washington, D.C., 1972.

Chapter 4

THE NATURE OF THE MATHEMATICAL LITERATURE

The current mathematical literature may be defined as that which forms the content of upper-level collegiate and graduate education and which represents present-day research interests. Mathematics that is studied in elementary and secondary school and in the first two years of college is a consolidation of results and ideas from earlier periods of mathematical history and, therefore, does not satisfy the definition of current mathematics.

I. FORMS

The current mathematical literature is chiefly serial in nature. The principal medium for the dissemination of mathematical information is the journal, followed by the book series, including series of monographs, proceedings of symposia, and translation series. These types of material are referred to as the "primary literature."

The least common form of publication is the nonserial book. Referred to as the "secondary literature," this category includes dictionaries, handbooks, mathematical tables, directories, histories, collected papers of individual mathematicians, and festschriften.

II. SUBJECTS

The broad headings used to classify papers and books in Mathematical Reviews gives an indication of the subject coverage of the literature. (A list of these headings appears on the cover of each

issue of Mathematical Reviews.) The literature reflects a wide
variety of research activity, ranging from traditional topics, such
as geometry, to more recent topics, such as homological algebra.
The increasing use of mathematics by other disciplines is evidenced
by literature dealing with the biological and behavioral sciences,
in addition to the traditional applications in the physical sciences.

III. LANGUAGES

While a reading knowledge of French, German, Italian, and Russian is useful to the mathematician, English has recently become
the dominant language in world mathematical circles. This can be
seen in the German journal Mathematische Annalen, where the percent-
age of papers in English rose from approximately 5 percent in the
mid-1930s to nearly 20 percent in the mid-1950s, and to 55 percent
in the mid-1960s [1].

Most countries with an academic tradition produce at least one
mathematical journal that is internationally recognized. English
language abstracts of many articles from these journals appear in
Mathematical Reviews. (A small number of abstracts in Mathematical
Reviews are written in French, German, and Italian.)

The American Mathematical Society has an active translation
program, notably in the area of Soviet mathematics. Many Russian
journals are now available in cover-to-cover translation. A list
of Russian journals in translation appears in every index issue of
Mathematical Reviews.

IV. SOURCES

Mathematical literature is issued by three general categories
of publishers: societies; universities and academies; and commercial
houses. All three groups issue both serial and nonserial publications,
covering both basic research and applied mathematics. It is not the
case in mathematics, as in many subjects, that research is published
by institutions and practical applications of the research are

reported by the commercial press. There are instances where mathematical societies place their publications in the hands of commercial houses, but this is for the purpose of administrative convenience. Bell of London, for example, is the publisher for the Mathematical Association.

There are few publications issued by the United States government that are of value to the mathematician. The only such publication abstracted in Mathematical Reviews is The Journal of Research of the National Bureau of Standards. Section B: The Mathematical Sciences. The National Science Foundation does not publish mathematics under its own name, but gives financial support to other mathematical publications, especially for translations of foreign-language works [2]. An interesting example of such cooperative publishing is a three-volume survey of mathematics that was translated from the Russian: Mathematics: Its Content, Methods, and Meaning, edited by A. D. Aleksandrov, A. N. Kolmogorov, and M. A. Lavrent'ev, and translated by S. H. Gould and T. Bartha, Russian edition 1956, American edition 1963. A National Science Foundation grant enabled the American Mathematical Society to translate and publish this work, which subsequently received wider circulation by being reissued by the M.I.T. Press.

REFERENCES

1. Committee on Support of Research in the Mathematical Sciences, Report of the Committee, The Mathematical Sciences. National Academy of Sciences, Washington, D.C., 1968.

2. J. A. Pemberton, How to Find Out in Mathematics, 2nd revised edition, Pergamon, Oxford, 1970.

Chapter 5

INFORMATION NEEDS IN MATHEMATICS

The research mathematician, the user of mathematics, and the student
of mathematics may each approach the mathematical literature in
various ways at various times for various purposes. It has been
demonstrated that it is easier to understand the pattern of use (and
the pattern of the information sources) by grouping the various kinds
of information needs into a series of "approaches" [1].

I. THE CURRENT APPROACH

This approach arises from a need to keep up to date with the
current progress in a field; to find out what other people in the
field are doing. Surveys have shown that in all subject fields the
most important source of such information is a fairly small core of
primary journals, systematically scanned. In most subject fields,
abstracting journals are generally regarded as secondary sources;
that is, they are used to locate and identify original papers [2].
In mathematics, because of the nature of the abstracting journals,
they can often be used as substitutes for the originals. Systematic
scanning of abstracting journals is an equally important way of
keeping up to date.

As the volume of literature has steadily increased, however,
the primary journals and the abstracting journals have found it
difficult to remain current. Information on the backlog of papers
for research journals is published in the February and August issues
of the Notices of the American Mathematical Society. The figures

given in the February 1977 issue show a range of three months (Journal of the American Statistical Association) to thirty months (Illi-
nois Journal of Mathematics) estimated time for a paper submitted
currently to be published [3]. As was noted earlier, the number of
abstracts in Mathematical Reviews has been doubling every eight or
ten years and presently there is a time lag of about twenty-four
months before a paper is reviewed. This forces the active researcher
to depend largely on personal communications among members of "in-
visible colleges"; that is, other researchers in his field of interest. Such communication occurs through conversations with colleagues,
attendance at meetings and symposia, correspondence, and the exchange
of preprints or preliminary drafts of papers.

Some steps have been taken to alleviate this problem of the
publication backlog.

The Bulletin of the American Mathematical Society contains a
"Research Announcements" section, the purpose of which is "to provide
early announcement of new results of such immediate importance to a
substantial set of mathematicians that their communication to the
mathematical community cannot await the normal time of publication"
[4].

Zentralblatt für Mathematik und ihre Grenzgebiete, the German
abstracting journal, has improved its currency by using authors'
abstracts instead of outside reviews for many of the research papers.

In addition, two current awareness services are presently available to mathematicians.

The first service, the Mathematical Title Service, provides
lists of titles in all areas of mathematics, based on detailed interest profiles of the subscribers. This service originally started
in 1968 as an experiment in automatic dissemination of information,
then called the Mathematical Offprint Service, which supplied off-
prints of papers. Papers which satisfied the subscriber's interest
profile were mailed to him on a continuing basis, along with title
listings of articles in which he indicated a peripheral interest.
However, this service was discontinued because it could not be made
self-supporting, and it was replaced by the Mathematical Title Service
[5].

The second service is in the form of a periodical, <u>Current</u>
<u>Mathematical Publications</u>, which is published biweekly by the Amer-
ican Mathematical Society. Each issue contains a list of the materials (papers, books, and other separately published items) that
have been received in the editorial offices of <u>Mathematical Reviews</u>
during a specific period, usually two weeks in length, and that will
eventually be reviewed. It is basically a classified index, arranged
according to the subject scheme used to classify entries in <u>Mathe-</u>
<u>matical Reviews</u>.

II. THE EVERYDAY APPROACH

This approach arises from the need for some specific piece of
information to be used in the course of daily work. For example,
the information needed may be a definition, an equation, a method
for solving a problem, or the correct spelling of a mathematician's
name. Such information can usually be found in a number of places.
Since the emphasis is on finding what is wanted as quickly as possible, the source is of little concern, provided it is reliable.

The quickest and easiest sources to consult are secondary sources,
such as encyclopedias, tables, dictionaries, handbooks, directories,
etc. These types of reference books have been specifically compiled
and arranged to furnish just such everyday information and they
usually provide what the searcher wants without making him look
beyond the book in hand.

Other secondary sources, such as abstracts and indexes are less
satisfactory. Either they provide more information on a topic than
is wanted, or, in many cases, they do not provide actual subject data,
but merely give references to other publications.

Unfortunately, there are relatively few mathematical reference
books that answer day-to-day specifics. The searcher is often required to locate such information in treatises and textbooks. Unless
he knows exactly where to look, this can be a time-consuming effort,
since such sources are not organized for quick "look-up" [6].

III. THE BACKGROUND APPROACH

The background approach may be considered a variation of the
everyday approach. This occurs in the course of daily work, although
not as a need for a specific piece of information, but for an outline account sufficient to understand a new or unfamiliar subject.

This need is fairly infrequent among research mathematicians,
but quite common among applied mathematicians who must familiarize
themselves with other fields. This approach is also common with
teachers and students who do not have time to digest original material [2].

Secondary sources, such as review journals and annual state-of-
the-art reports are the types of literature that meet this need in
other sciences. Mathematics, however, is less susceptible to rapid
review than any of the other sciences. Because of the highly abstract character and extreme specialization of much of current mathematics, annual reviews of its progress are hardly feasible. A technical understanding of the field in which the advances lie is necessary to understand the significance of most mathematical progress.
For these reasons, review material appears less frequently than in
other disciplines and usually in the form of a book, rather than as
a review journal or article. One notable exception is Ergebnisse
der Mathematik und ihrer Grenzgebiete, Springer-Verlag, Berlin, which
is discussed in Chapter 10 [7].

IV. THE EXHAUSTIVE APPROACH

This approach, which occurs less frequently than the current
or everyday approaches, involves searching the literature for all
relevant information on a given subject. "Exhaustive" is, of course,
a relative term, dependent upon the purpose of the search, and the
nature of the subject in question.

A literature search usually begins with secondary sources, such
as abstracts and indexes, starting with the most recent issue and
progressing back in time through the earlier volumes. Once the pri-

mary sources have been identified and located, they themselves are used in the search by following the references cited at the end of each original article or book. Once these references have been located, their references can be followed, and so on. This is a common procedure in conducting a literature search.

Other secondary sources, such as handbooks and encyclopedias, often contain additional references and should not be overlooked.

Since there are very few separately published subject bibliographies in mathematics, it is usually necessary to consider treatises and monographs as reference tools. The bibliographies contained in these types of books are always extensive and are often the only bibliographies available on a given topic.

REFERENCES

1. M. J. Voigt, Scientists' Approaches to Information, American Library Association, Chicago, 1961.

2. D. J. Grogan, Science and Technology: An Introduction to the Literature, Archon Books, Hamden, Conn., 1970.

3. Notices of the American Mathematical Society, 24, 118, (1977).

4. Bulletin of the American Mathematical Society, 83, 119, (1977).

5. Conference Board of the Mathematical Sciences, Proceedings of a Conference on a National Information System in the Mathematical Sciences, Washington, D.C., 1970.

6. K. O. May, "Problems of Information Retrieval in Mathematics," Proceedings of the Canadian Mathematical Congress, Winnipeg, 1971, pp. 477-484.

7. N. G. Parke, Guide to the Literature of Mathematics and Physics, 2nd edition, Dover, New York, 1958.

Chapter 6

THE ORGANIZATION OF MATHEMATICAL LITERATURE IN LIBRARIES

The library serves the mathematician as the laboratory serves the
experimental scientist. No scientist would begin working in a laboratory without some instruction in the use of the equipment. But
the mathematician rarely receives any training in the use of his
equipment, the library collection. A general understanding of the
organization of libraries, what they contain, and the services they
offer is essential for fully exploiting the mathematical literature.

The contents of a library are organized by means of two devices:
(1) a classification scheme, which provides for the physical arrangement of materials on the shelves; and (2) a catalog, which provides
a record of the materials in the collection and their locations.

I. LIBRARY CLASSIFICATION

The purpose of library classification is to organize the collection for use. To achieve this, books are placed on the shelves according to subject, and are arranged alphabetically by author within
each subject group. To assure that they will always be found in the
same relative position, numbers are assigned. These numbers, called
"call numbers," consist of two parts: the class, or subject number;
and the book number. The class number will be the same for all books
dealing with the same subject. The book number is unique to each
book, distinguishing it from other books having the same class number.
Periodicals may be treated in the same manner as books. Generally,

however, it is the practice not to classify periodicals, but to
shelve them by title in a separate location.

Many different classification schemes exist. The two that are
in general use in American libraries, the Library of Congress Classification and the Dewey Decimal Classification, are the ones discussed
here.

A. Library of Congress Classification

This system, which was developed in the early years of this
century for the collections of the Library of Congress, has been
adopted by most college and university libraries in the United States.

The scheme consists of over thirty separately published volumes
or "schedules" for the main classes and subclasses. The individual
schedules are revised periodically and the one which contains mathematics is now in its sixth edition.

The notation of the scheme is "mixed"; that is, it consists of
letters and numbers. The main classes are designated by single capital letters (e.g., "Q" stands for science), the subclasses by two
capital letters (e.g., "QA" stands for mathematics), and the divisions and subdivisions of the subclasses are designated by integral
numbers in ordinary sequence ranging from 1 to 9999 (e.g., 372 stands
for ordinary differential equations). A complete Library of Congress
call number consists of this class notation, plus an author notation,
which is an initial letter followed by one or two Arabic numerals.
For example, An Introduction to Ordinary Differential Equations, by
G. J. Etgen has the following call number:

QA.........The subclass for mathematics

372........The subdivision by integral numbers meaning
           ordinary differential equations.

E8........The author number. Etgen.

An outline of the QA subclass is given in Table 1.

B. Dewey Decimal Classification

This system, first introduced in 1876, was used in American
libraries of all types prior to the development of the Library of
Congress Classification. Today, it is used mainly by public libraries. However, it, and its descendent, the Universal Decimal Classification, are still the most widely used systems in libraries outside
the United States. Also, many printed bibliographies (e.g., American
Book Publishing Record, Bowker, New York) are arranged by the Dewey
Classification.

The universal popularity of this system, which is now in its
eighteenth edition, has been attributed to the simplicity of its
structure and to the fact that it uses only numbers in its notational
scheme.

The Dewey Decimal Classification divides all knowledge into nine
main classes, plus a generalia class, through a three figure array
of main classes, as follows:

000	Generalia	500	Pure Science
100	Philosophy	600	Applied Science
200	Religion	700	Fine Arts
300	Social Sciences	800	Literature
400	Language	900	History and Geography

Each main class is divided into ten subclasses, with the first section covering the general works of the entire class:

500	Pure Science	520	Astronomy
510	Mathematics	530	Physics, etc.

Each subclass is divided into sections and, again, the first section
covers the general works of the entire subclass:

510	Mathematics
511	Generalities (including Logic)
512	Algebra, etc.

A decimal point is used after the third number and, in theory, expansion can be carried out indefinately to any degree of specificity.

An outline of the mathematics subclass is given in Table 2.

TABLE 1

Outline of Subclass QA

QA 1-99 General Mathematics

  1-7 [General Periodicals, Reference Books, Collected Works]

  8-10 [Philosophy and Logic, including Set Theory]

  11-20 Study and Research

  21-35 History [to 20th Century]

  36-85 General Works [Textbooks, Handbooks, Tables Computers]

  90 Graphic Methods (General)

  93-95 Popular Works, Mathematical Recreations, Miscellany and Curiosa.

  101-145 Elementary Mathematics. Arithmetic

  101-107 Textbooks

  109-119 Special Aspects of the Subject as a Whole

  135-139 Study and Teaching

  141 Numeration, Number Concept, Numeration Systems

  145 Arithmetic and Algebra

TABLE 1 (continued)

QA   150-271   Algebra
               150-161   [Periodicals, Textbooks, General Works]
               162-271   [Special Topics, e.g., Homological Algebra]
               273-274   Probabilities
               276-280   Mathematical Statistics

        281   Interpolation

        292   Sequences

        295   Series

    297-299   Numerical Analysis

    300-433   Analysis
               300-302   [General Works]
               303-316   Calculus
               320-433   [Special Topics, e.g., Theory of Functions]

    440-699   Geometry
               440-447   [General Works]
               451-469   Elementary Geometry
               471-699   [Special Topics, e.g., Projective Geometry,
                         Topology, etc.]

    801-939   Analytic Mechanics [Classical Applied Mathematics]
               801-809   [General Works]
               821-871   [Mechanics of Particles and Systems]
               901-930   Statics and Dynamics of Fluids
               931-939   Elasticity, Plasticity

TABLE 2
Outline of Subclass 510

510 Mathematics
511 Generalities
 .2 Inductive and Intuitive Mathematics
 .3 Symbolic Logic
 .4 Approximations and Expansions
 .5 Theory and Construction Graphs
 .6 Combinatorial Analysis
 .7 Numerical Analysis
 .8 Mathematical Models
512 Algebra
 .1 Algebra Combined with Other Branches of Mathematics
 .2 Groups
 .3 Fields
 .4 Rings, Integral Domains, Ideals
 .5 Linear, Multilinear, Multidimensional Algebras
 .7 Number Theory
 .9 Pedagogical Algebra
513 Arithmetic
514 Topology
 .2 Algebraic Topology
 .3 Topology of Spaces
 .7 Analytic Topologies
515 Analysis
 .1 Analysis and Calculus Combined with Other Branches
 .2 Generalities
 .3 Differential Calculus and Equations
 .4 Integral Calculus and Equations
 .5 Special Functions
 .6 Other Analytic Methods
 .7 Functional Analysis
 .8 Functions of Real Variables
 .9 Functions of Complex Variables

TABLE 2 (continued)

516 Geometry

.1 Generalities

.2 Euclidean Geometry

.3 Analytic Geometry

.4 Affine Geometry

.5 Projective Geometry

.6 Abstract Descriptive Geometry

.9 Non-Euclidean Geometries

517 [not assigned]

518 [not assigned]

519 Probabilities and Applied Mathematics

.2 Probabilities

.3 Game Theory

.4 Applied Numerical Analysis

.5 Statistical Mathematics

.7 Programming

.8 Special Topics

No specialist is ever satisfied with someone else's classification of his subject field. Mathematicians have been particularly
outspoken in criticizing library classification schemes for failing
to meet the needs of contemporary mathematics. This criticism has
validity, since both the Dewey and Library of Congress systems are
based on the mathematics of the nineteenth century; that is, they
are organized around the traditional divisions of arithmetic, algebra, geometry, and analysis. When new topics are introduced into
the systems they have to be accommodated within this existing framework.

Furthermore, the basic arrangement of both schemes is by discipline and various aspects of a given subject may appear in any number of disciplines. For example, applications of mathematics to

other disciplines are not found in QA or 510, but are classed with
those disciplines to which they are applied.

For these reasons some books will always be classed in the
"wrong" places, from individuals' points of view and interests.
Browsing in the library stacks must be supplemented by consulting
the library catalog if all pertinent material is to be located.

II. LIBRARY CATALOGS

The catalog of a library is the key to its holdings. The importance of understanding how to use the catalog cannot be stressed too
strongly. Very often, a great deal of material is missed by avoiding
the catalog and by relying solely on the shelf arrangement of the
materials. Furthermore, because of the bibliographic information
it provides, the catalog serves as an important reference tool. It
is often the most convenient source for verifying authors' names,
titles of publications, dates of publication, etc.

Catalogs may appear in various forms. Prior to the late nineteenth century the book catalog was common, either in the form of a
handwritten or printed book or loose-leaf sheets locked in binders.
With the development of photographic and computer techniques of book
production there has been a resurgence of interest in book catalogs
in recent years. Some libraries have also experimented with catalogs
in microform, either reels of microfilm or sheets of microfiche. But
the most frequently encountered form of catalog, and the one described
here, is the card catalog: sequence of index cards, approximately
3 X 5 inches, filed in drawers.

Practically every document in the library will be represented
in the catalog by two or more cards (or entries, in library terminology). The most common ones are for author, title, and subject;
but when there is occasion for them, entries are made for joint
author, editor, translator, and for series.

The Author Entry, so called because the author's name appears
on the top line, contains the following information: author's name

in full; title; imprint (i.e., place of publication, name of publisher, and date of publication); number of pages (or volumes if a
set); and series note. A statement of contents, or other types of
information are often appended when appropriate.

The Title Entry is a duplicate of the Author Entry, except that
the title appears on the top line of the card.

The Subject Entry is also a duplicate of the Author Entry, but
in this instance the subject appears on the top line. (The subject
is usually typed or printed in black capitals to make it visually
distinctive.) Subject entries are made for as many topics as are
discussed comprehensively in the work. Uniformity of subject entry
is achieved through the use of selected terms, called subject headings, prepared for the purpose. The standard list in college and
university libraries in the United States is Subject Headings Used
in the Dictionary Catalogs of the Library of Congress. A copy of
this list is usually located near the card catalog and should be consulted before beginning a subject search if one is not familiar with
the terminology of library subject headings.

In addition to these three basic types of entries, other entries
may be made for joint authors, editors, and titles of series, as
noted above. Such entries also contain the full information of the
Author Entry.

A Dictionary Catalog is one in which the cards for all entries
are filed in one alphabetical sequence. A Divided Catalog, as its
name implies, is one in which the cards are divided into two sequences: one for subject entries and one for author, title, and
other nonsubject entries. For convenience in filing and searching
most libraries have adopted the Divided Catalog and a few even have
a three-way division of subject, title, and names.

For periodicals and other serial publications, supplementary
lists are usually maintained. The catalog entries for these kinds
of materials will give minimal information, such as the date the
library began subscribing to a journal, and the supplementary list
will provide full information about the library's holdings.

One final, and vital, piece of information provided by the catalog is the location of documents in the library. The call number, which appears in the upper left-hand corner of each card, will indicate whether the item is shelved in the book stacks or in a special location. Most libraries have special sections for reference books, periodicals, and microforms.

Chapter 7

JOURNALS

I. GENERAL

The first scientific journal is generally acknowledged to be the
Journal des Scavans, founded in 1665 by de Sallo, a counsellor of
the French court of parliament. Its avowed purpose was "to catalog
and give useful information on books published in Europe and to
summarize their works, to make known experiments in physics, chemistry, and anatomy that may serve to explain natural phenomena, to
describe useful or curious inventions or machines and to record
meteorological data, to cite the principal decisions of civil and
religious courts and censures of universities, to transmit to readers
all current events worthy of the curiosity of men." De Sallo claimed
that his journal was instituted "for the relief of those either too
indolent or too occupied to read whole books" [quoted in 1]. In
1816 it became the Journal des Savants, still a leading journal, but
now of a literary nature.

Only four months after de Sallo's Journal began, the first
English scientific journal appeared. The Philosophical Transactions,
published by members of the Royal Society of London, recorded experiments conducted by members of the Society and correspondence with
their European counterparts. It did not include legal or theological
questions. The journal became the official organ of the Royal Society of London with Volume 47 and is still in existence.

Two scientific journals published in Germany and patterned after
de Sallo's Journal were Miscellanea Curiosa (1670-1705) and Acta

49

Eruditorum (1682-1731). Other significant early journals were the
Acta Medica et Philosophia Hafneinsia, which appeared in Copenhagen
from 1673-1680, the Raccolta d'Opuscoli Scientifici e Fililogici,
published in Venice from 1728-1757 and continued as the Nuova Raccolta Opuscoli Scientifici from 1755 to 1787, Le Pour et le Contre,
published in Paris from 1733 to 1740, and the Göttingische Zeitung
von Gelehrten Sachen, issued from 1739 to 1752, but continued under
varying titles well into the nineteenth century [1].

One characteristic of these early journals was that they were
general in nature, covering several branches of science. It was not
until the latter part of the eighteenth century that specialized
journals appeared, notably in the field of chemistry. The first
journal devoted solely to mathematics was the Annales de Mathéma-
tiques Pures et Appliquées (France, 1810-1831). The Journal für die
Reine und Angewandte Mathematik, founded in Germany in 1826, and
usually referred to as Crelle's Journal, after its founder, is the
oldest mathematics journal still in existence.

As was the case with the early general scientific journals, the
first specialized journals were of the digest type of publication
and were published by individuals rather than under the auspices of
learned societies.

With the establishment of national societies in the nineteenth
century, the number of journals devoted to mathematics increased.
As the rate of mathematical research has increased and become more
specialized in this century, the journals have also increased in
number and become more restricted in their subject scope. At present,
general and specialized mathematical journals are published by national societies, universities, and commercial publishers.

II. NATIONAL SOCIETIES

With the exception of the Wiskundig Genootschap, founded in
Amsterdam in 1778, most national societies did not appear until the
second half of the nineteenth century.

The oldest of the major national societies is the Moskovskoe Matematicheskoe Obshchestvo (Moskow Mathematical Society), organized in 1864. It publishes the journal Matematicheskii sbornik (1865-) and Trudy Moskovskogo Matematicheskogo Obshchestva (Transactions of the Moscow Mathematical Society, 1952-).

The London Mathematical Society, founded in 1865, is the national mathematical society of England. It publishes Proceedings (1865-), a Journal (1926-), and a Bulletin (1969-).

The Edinburgh Mathematical Society was founded in 1883 and publishes Proceedings (1883-) and Mathematical Notes (1909-).

The Société Mathématique de France was founded in 1872 and publishes a Bulletin (1873-).

In Italy the Circolo Matematico di Palermo, founded in 1884, published Rendiconti (1887-), which for many years was one of the most important mathematical journals in the world. It is currently published by the Societa Industrie Riunite Editorial, Palermo. The Unione Matematica Italiana, established in 1922, is now the national society of Italy and publishes a Bollettino (1922-).

The national society of Germany is the Deutsche Mathematiker-Vereinigung, founded in 1890. It publishes Jahresbericht (1890-).

Today most countries have national societies which publish at least one journal. Two of the most important societies outside Europe and the Americas are the Indian Mathematical Society, which began in 1907, and the Mathematical Society of Japan, which was formed after World War II. Each publishes one journal [2].

National mathematical societies are concerned with the three major areas of research, education, and applications. In some countries all these areas are covered by one society, in other countries different societies deal with each separately. Most societies issue, free of charge, brochures describing their aims, activities, conditions of membership, and lists of their publications [3].

In the United States there are several professional societies, with different purposes. The Conference Board of the Mathematical Sciences is a consortium of several of these societies, whose consti-

tutional objective is to coordinate the activities of its member
societies. At present the Conference Board is comprised of twelve
societies, including the American Mathematical Society, a research-
oriented organization, The Mathematical Association of America, an
education-oriented organization, and several societies that are concerned with applications. The journals of the American Mathematical
Society and The Mathematical Association of America are discussed
below and the journals of some of the applied mathematical societies
are discussed in Chapters 13, 14, and 15.

A. The American Mathematical Society

The American Mathematical Society is one of the largest and most
important societies in the world. Founded as the New York Mathematical Society in 1888, it acquired its present name in 1894. Its
primary purpose is to preserve, supplement, and utilize mathematical
research. Its membership is drawn chiefly from the United States,
but it also includes members from Canada and other countries. The
Society not only issues journals under its own name, but it publishes
translations of foreign journals, it has cooperative agreements with
other mathematical societies, and it contributes editorial and financial support to a number of journals published by other organizations.

Of the five journals published under the Society's imprint,
three are devoted solely to the results of original research.

Transactions of the American Mathematical Society (1900-) and
Proceedings of the American Mathematical Society (1950-) are both
issued monthly and cover all areas of pure and applied mathematics.
Ordinarily, longer papers are published in the Transactions and
shorter ones in the Proceedings. A strict editor-referee system,
which insures the highest level of excellence of the papers, makes
these among the more prestigious journals in which to publish.

The third journal, which is devoted to research in a special
area, is Mathematics of Computation (formerly Mathematical Tables
and Other Aids to Computation, 1943-). This is a quarterly journal,

containing original papers in numerical analysis, the application of
numerical methods and high-speed calculator devices, the computation
of mathematical tables, the theory of high-speed calculating devices,
and other aids to computation. In addition, reviews and notes in
these and related fields are published.

Five journals of Russian translations are currently published
by the Society. Three are devoted to pure mathematics, one is specialized, and one covers all areas of pure and applied mathematics.

Soviet Mathematics - Doklady (bimonthly) is a translation journal containing the entire pure mathematics section of the Doklady
Akademii Nauk SSSR, the Reports of the Academy of Sciences in the
USSR. Doklady publishes five hundred articles a year, each about
four pages long. It provides a comprehensive up-to-date survey of
research in the USSR.

Mathematics of the USSR - Izvestija (bimonthly) is a cover-to-
cover translation of Izvestija Akademii Nauk SSSR Serija Matemati-
českaja, published by the Academy of Sciences of the USSR. This is
a journal of current research in all fields of pure mathematics.

Mathematics of the USSR - Sbornik (monthly) is a cover-to-cover
translation of Matematičeskii Sbornik (Novaja Serija), published by
the Moscow Mathematical Society and the Academy of Sciences of the
USSR. This is a journal of current research in all fields of pure
mathematics.

Theory of Probability and Mathematical Statistics is a cover-to-
cover translation of the Teorija Verojatnostei i Matematiceskaja
Statistika published by Kiev University, beginning with the 1970
Soviet publication.

Vestnik of the Leningrad University (Mathematics) is the complete translation of the mathematics section of the Vestnik Lenin-
gradskogo Universiteta, beginning with the Soviet publication of
1968. All fields of mathematics are covered.

Other Society translations, which are published as book series,
are discussed in Chapter 10.

B. The Mathematical Association of America

The Mathematical Association of America, founded in 1915, is another United States-based society which includes a large number of Canadians in its membership. The Association is concerned primarily with collegiate-level mathematics and teaching, and publishes three journals.

The American Mathematical Monthly (1894-), published ten times a year, contains original and expository articles at the undergraduate and beginning graduate level. Its departments of "Mathematical Notes," "Classroom Notes," and "Mathematical Education Notes" are of special interest to teachers of mathematics. The departments of "Elementary and Advanced Problems," "News and Notices," and "Reviews" are of even more general interest. "Official Reports and Communications" of the Association are published in another section of the Monthly. These include minutes of the meetings of the Association and its Sections and of the Board of Governors of the Association.

Mathematics Magazine (1926-), published five times a year, is similar in content to the Monthly, but focuses at a somewhat lower level. It includes expository articles, problems, reviews, and other features of general interest to undergraduate students and faculty.

The Two-Year College Mathematics Journal (1970-), published four times a year, is written and edited largely by two-year college faculty. This journal features mathematical and pedagogical articles, problems, book reviews, and other specialized departments of special interest to the two-year college community. It also contains news and notices of Association activities, members, and meetings similar to those published in the Monthly.

III. UNIVERSITY PUBLICATIONS

Societies are not the only publishers of mathematical journals; universities also produce them. In fact, the first mathematical journal in the United States was the American Journal of Mathematics, founded in 1878 by J. J. Sylvester when he was a professor at Johns

Hopkins University. As noted earlier, the journal is still in existence and is now published jointly by the University and the American
Mathematical Society. Currently, many universities with strong mathematics programs, in the United States and world wide, publish journals; some are general in nature and some are specialized. Some
examples of the variety of journals produced by university presses
are listed below.

Two specialized journals are the Journal of Differential Geom-
etry (1967-) published by Lehigh University and the Quarterly of
Applied Mathematics (1943-) published by Brown University. The
Rocky Mountain Journal of Mathematics (1971-) is a quarterly published by a consortium of western schools. The University of
Toronto (Canada) publishes two journals for other organizations.
They are the Canadian Journal of Mathematics/Journal Canadien de
Mathematique (1949-), the bimonthly journal of the Canadian Mathematical Congress, and Historia Mathematica (1974-), a quarterly
sponsored by the International Union for the History and Philosophy
of Science, Division of the History of Science, Commission on History of Mathematics.

In England, Oxford University publishes a general journal, the
Quarterly Journal of Mathematics (1930-), and the University of
Sheffield publishes a specialized one, The Journal of Applied Proba-
bility (1964-).

IV. COMMERCIAL PUBLISHERS

Until quite recently, commercial firms have played a more important role in journal publishing in Europe than in the United States.
For example, three important journals have for many years been published by Springer-Verlag of Berlin. They are Monatshefte für
Mathematik (1890-), Mathematische Zeitschrift (1918-), and Mathematische Annalen (1920-). Articles in the first journal are in
English and German, and in English, German, or French in the last
two.

North-Holland Publishing Company of Amsterdam, The Netherlands, publishes several journals, including: Artificial Intelligence (1970-), Discrete Mathematics (1971-), General Topology and Its Applications (1971-), and Journal of Pure and Applied Algebra (1971-).

North-Holland also publishes Annals of Mathematical Logic (1969-) for the Association for Symbolic Logic, and Mathematical Programming (1971-) for the Mathematical Programming Society. (Text in English, French, and German).

Pergamon Press in England has published Topology: An International Journal of Mathematics since 1962. This is a quarterly journal, with text in English, French, German, or Italian.

In 1960, Academic Press of New York inaugurated an impressive program of publishing journals devoted to special areas of mathematics. Currently it issues the following titles: Journal of Mathematical Analysis and Applications (1960-), Journal of Algebra (1964-), Journal of Differential Equations (1965-), Journal of Combinatorial Theory (1966-), Journal of Functional Analysis (1967-), Journal of Approximation Theory (1968-), Journal of Number Theory (1968-), and Journal of Multivariate Analysis (1971-).

Two other journals, both published by Marcel Dekker Inc., are Communications in Algebra (1974-), and Communications in Partial Differential Equations (1976-).

REFERENCES

1. Bernard Houghton, Scientific Periodicals: Their Historical Development, Characteristics, and Control, Linnet Books, Hamden, Conn., 1975.

2. "Mathematics, Societies of," in Encyclopaedia Britannica, Volume 14, 1973.

3. J. E. Pemberton, How to Find Out in Mathematics, 2nd edition, Pergamon Press, Oxford, 1969.

CHAPTER 8

ACCESS TO JOURNALS: BIBLIOGRAPHIES, INDEXES, ABSTRACTS

I. BIBLIOGRAPHIES

The first step in using journals is to identify and locate them.
The various types of bibliographies that have been compiled for that
purpose is the subject of the first section of this chapter. Bibliographic control of the contents of journals through indexes and
abstracts is discussed in the remaining two sections. Individual
mathematical abstracting journals are described in Chapter 9.

A. Current Bibliographies

The most comprehensive and useful guide to journals currently
being published throughout the world is Ulrich's International Periodicals Directory (Bowker, New York). The 16th edition, dated
1975/76, is in two volumes and contains entries for 57,000 in-print
journals. The entries are arranged alphabetically by title within
subject groupings (e.g., Astronomy, Computer Technology and Applications, Mathematics, Statistics, etc.). Detailed information for
each title includes: date of its origin; frequency of issue; price;
name and address of publisher; language(s) of text; presence of
illustrations, reviews, or bibliographies; coverage by abstracting
and indexing services; and circulation figures. The main section is
supplemented by a title and specific subject index and a list of
journals that have ceased publication since 1973/74, the date of the
15th edition.

A useful companion volume to Ulrich is Irregular Serials and
Annuals, 4th edition, 1976/77. This is a listing of approximately
30,000 yearbooks, annual reviews, advances in, and similar publications .that are issued annually, or less frequently than once a year,
or at irregular intervals. Again, the main sequence is by subject
groupings, with complete bibliographic information for each title,
and a supplementary title and specific subject index.

Ulrich and Irregular Serials and Annuals are published biennially, in alternating years. A third publication, the Bowker's Serials
Bibliography Supplement, is issued intermittently to extend their
coverage between editions. The latest edition of this title, dated
1976, includes about 7,100 current serials.

National listings of journals published within a particular
country provide another source of information about current journals.
Often a general bibliography of journals and a more specialized listing that covers only science and technology will be available for
a country. However, there is little difficulty in selecting mathematical journals, since such lists are usually arranged by subject
groupings.

An example of a general bibliography is David Woodworth, Guide
to Current British Journals, 2nd edition, Library Association,
London, 1973. This work is in 2 volumes, the first volume containing
approximately 4,700 current titles, arranged by the Universal Decimal
Classification. Appendixes list journals that contain abstracts,
and names of professional societies and their publications. The
second volume is a directory of British publishers and titles of
their journals.

An example of a more specialized listing is the Directory of
Canadian Scientific and Technical Periodicals, which has been published biennially since 1961 by the National Science Library of
Canada. This is a classified listing of about 900 current titles in
science, technology, and medicine.

Current titles may also be located through holdings lists of
major scientific and technological libraries; for example, National

Lending Library for Science and Technology, Current Serials Received, Her Majesty's Stationery Office, 1967. In the United States, the Science and Technology Division of the Library of Congress has published A List of Scientific and Technical Serials Currently Received by the Library of Congress, Government Printing Office, 1960. Such lists are usually arranged in alphabetical title sequences, so their chief value lies in bibliographical checking and location of titles, rather than as subject lists.

Abstracting and indexing services can also be used to identify titles of journals currently being published throughout the world in the mathematical sciences. A list of journals covered by an abstracting journal is normally published in each issue, or if the coverage is extensive it may be published at the end of a volume. In the case of Mathematical Reviews, for example, the list is included in the index to each volume.

B. Union Lists

Directories of the combined journal holdings of several libraries are known as union lists. Union lists may be local or national, general or specialized, but in all cases their chief purpose is to facilitate the locations of individual titles. Union lists are invaluable in retrospective searching since their coverage of older material is excellent. They are the major tools for tracing locations for sets of a particular title, and they are also of value in checking bibliographical details such as changes in titles or verifying places of publication. They are of little relevance, however, in subject selection as they are alphabetical title listings of journals with details of holdings of the cooperating libraries.

The Union List of Serials in Libraries of the United States and Canada, 5 volumes, 3rd edition, Wilson, New York, 1965, lists 150,000 titles published up to 1949 held in the Library of Congress and other cooperating libraries. No further editions are planned, but since 1950 the work has been continued as New Serial Titles, a monthly pub-

lication prepared by the Library of Congress. This has annual,
quinquennial and decennial cumulations.

The British Union Catalogue of Periodicals, 4 volumes, Butterworths, London, 1955-58, Supplement 1960, lists 140,000 titles held
in 440 British libraries. Quarterly supplements listing new titles
in the cooperating libraries and titles published after 1960 have
been issued since 1964. These cumulate annually and quinquennially.
A separate list of titles in science and technology is now published
each year as the World List of Scientific Periodicals and is described
below.

The Union List and the British Union Catalogue are general in
scope; they cover journals in all subject fields. An example of a
more specialized union catalog is the World List of Scientific Periodicals Published in the Years 1900-1960, 3 volumes, 4th edition,
Butterworths, London; Archon Books, Hamden, Conn., 1963-66. This
lists 60,000 titles with the holdings of 300 libraries. The main
work is supplemented by New Periodical Titles, 1960-1968, Butterworths, London, 1970. No further editions will appear, but the work
is being continued in the annual supplements of the British Union
Catalogue of Periodicals as the World List of Scientific Periodicals;
Scientific, Medical and Technical Entries from the British Union
Catalogue of Periodicals. The World List of Periodicals does indeed
live up to its name in including periodicals from all over the world.
But one must keep in mind that it is a national and not a world list,
since it only indicates holdings in British libraries.

C. Retrospective Bibliographies

As mentioned above, union lists are useful in retrospective
searching since they include older journals and they record all
variations in titles and other bibliographic data that have occurred
over the years. There are also two general bibliographies of early
scientific periodicals that are important for retrospective or exhaustive searching.

H. C. Bolton, Catalogue of Scientific and Technical Periodicals, 1665-1895, 2nd edition, Smithsonian Institution, Washington, D.C., 1897. This is an alphabetical title catalog of almost 9,000 titles, which its preface states "is intended to contain the principal independent periodicals of every branch of pure and applied science, published in all countries from the rise of this literature to the present time." A supplementary volume to Bolton is:

S. H. Scudder, Catalogue of Scientific Serials of all the Natural, Physical and Mathematical Sciences, 1633-1876, Harvard University, Cambridge, Mass., 1879. The 4,400 titles in this catalog are arranged by country, then by place of publication, with indexes of town, title, and subject. It excludes technology, but includes the transactions of learned societies (omitted by Bolton).

As the foregoing discussion has indicated, bibliographies of journals fall into the following categories: journals currently being published world-wide; journals currently being published within a given country; holdings lists of individual libraries; union lists; and, retrospective lists. Within each of these categories are found bibliographies covering journals in all subject fields; journals in the fields of science and technology; and, journals of an individual subject field. A useful guide to special subject fields within science and technology (a bibliography of bibliographies, in fact) is M. J. Fowler, Guides to Scientific Periodicals: An Annotated Bibliography, Library Association, London, 1966. This work covers 1,060 publications and is arranged in classified order within three main sections: (1) Universal guides (those covering journals of all countries): (a) general, (b) special subject lists; (2) Guides to the journals published by international organizations; and, (3) National and other regional guides. There is also an author/title/subject index.

It is obviously impossible within the scope of this book to discuss all of the available guides, lists, catalogs, and bibliographies that aid in the identification and location of journals. The reader interested in more information on this topic is referred to the two excellent books listed at the end of this chapter.

II. INDEXES

Once journals have become back files or bound volumes, the job
of locating specific information in past issues is made less tedious
by the use of indexes. Most journals provide an index to their contents, commonly once a year, or once per volume, if this is not
yearly. It may be included in the final issue of the year (or volume) or published separately. Since each journal is a law unto itself in this matter, Ulrich's International Periodicals Directory,
which indicates the indexing practices of individual journals, is a
helpful source [1].

Some publishers issue cumulative indexes covering five, ten, or
even more years of a journal. Again, practices vary and Ulrich
should be consulted for this information.

Even more valuable are indexes that analyze the contents of
several journals. An example of this is the Current Index to Statistics, which covers the statistical journal literature and which
is more fully described in Chapter 14.

Unfortunately, there is no comprehensive index of this type
for the journal literature of all the mathematical sciences. There
are a few cumulative indexes to individual publications; for example,
Mathematical Reviews Cumulative Author Indexes, and Index to Translations Selected by the American Mathematical Society. These indexes
are discussed in connection with Mathematical Reviews in Chapter 9
and American Mathematical Society Translation Series in Chapter 10,
respectively.

III. ABSTRACTS

Abstracts can be traced back to the very beginnings of the
journal literature of science and technology. Many of the early
journals performed an abstracting function in their efforts to digest the significant contemporary developments. But it was not until
the appearance of the specialized journal that the specialized abstracting journal as we know it today emerged. Today, as then, the

abstract is probably the most convenient method of summarizing and
recording for both current and future use the information contained
in the journal literature [2].

An abstract may be broadly defined as a summary of the information in a document, accompanied by an adequate bibliographic citation
to enable the document to be traced. This citation is of vital
importance in that it acts as the link between the secondary source
(the abstract) and the primary source (the original paper). The
reader, therefore, should be careful to note all the items in a
citation: (1) title and author(s) of the original paper; (2) title
of the journal in which the original paper appears; (3) volume and
part number of the journal; (4) date of issue; and, (5) pages over
which the paper extends.

The common practice of abbreviating journal titles in citations
can often be a problem, particularly when the abbreviations are not
self-evident. The abbreviations of journal titles used by abstracting services are not usually made according to official national
or international standards. An interesting account of this development may be found in [2]. It is sufficient to note here that a list
of abbreviations will be provided somewhere in the abstracting journal; often in the index issue, if the list is extensive.

With respect to the second element of an abstract, the summary,
a distinction is commonly drawn between an "indicative" abstract and
an "informative" abstract. The former, as the name implies, gives a
broad indication of the scope and content of the original paper; it
enables the reader to decide whether he wants to see the original.
The informative abstract summarizes the principal arguments and gives
the significant methods and data of the original paper. It can sometimes serve as a substitute for the original, expecially if the
original is not available or is written in a foreign language. When
an abstract contains an evaluation of the original it is properly
called a review.

Methods of producing abstracts vary among the different services.
This is reflected in the type of abstract and the degree of abstracting that results and is a point to keep in mind when using an abstract-

ing service. Some abstracting services employ staffs of professional
abstracters; these abstracts are usually indicative. Other services,
such as Mathematical Reviews, send the original papers to subject
specialists to be abstracted. The resulting abstracts are of the
informative type and are often critical, as well. An abstract written
by the author of the original is also informative, although not likely
to be critical (i.e., objective). The occurrence of author abstracts
has increased greatly in recent years, primarily as a means of re-
ducing delays in the publication of the abstracts.

Abstracts are produced by different kinds of organizations:
professional societies; commercial publishers; and, governments.

Most abstracting publications in the mathematical sciences are
produced by learned societies and professional bodies. For example,
Mathematical Reviews is published by the American Mathematical Society, and Applied Mechanics Reviews, by the American Society of Mechanical Engineers.

Abstract compilation is so laborious and unprofitable a task
that it does not attract many commercial publishers, but there are
exceptions. Springer-Verlag is the publisher of Zentralblatt für
Mathematik und ihre Grenzgebiete.

As noted earlier, the United States government publishes little
mathematics under its own name, but is involved indirectly by sub-
sidizing both primary and secondary publications of institutions and
societies. One exception is the Journal of Research of the National
Bureau of Standards. Section B: The Mathematical Sciences. Each
issue of this journal contains abstracts of the Bureau's publications.

In some countries the state itself operates specialized abstracting organizations. In France the Centre National de la Recherche
Scientifique compiles Bulletin Signaletique. This is published in
several sections, the first of which is entitled Mathematiques. In
the Soviet Union, Referativnyi Zhurnal, compiled by VINITI (All-
Union Institute of Scientific and Technical Information), contains
sections devoted to the mathematical sciences [1].

The most common form for an abstracting service is a journal
devoted to abstracts. Some abstracts may appear as a feature within

a primary journal, but this is not typical in mathematics. The major use of abstracting services is in documenting journal articles, although some cover non-journal publications, as well.

Abstracting journals help the reader to maximize the use of the time he has available for reading. They bring together and identify papers relating to any given subject that are distributed over a wide range of journals, thus eliminating the need to scan the individual journals. This is the current awareness use of abstracting services.

Retrospective search is another major use of abstracting services. The retrospective accumulation of papers in the journals covered by the abstract journal can be accessed by using the subject indexes and author indexes appended to most abstract journals. Without these indexes the reader would be faced with the task of scanning the numerous individual indexes of the primary journals covering the subject.

REFERENCES

1. Denis Grogan, Science and Technology: An Introduction to the Literature, 2nd edition, Linnet Books, Hamden, Conn., 1973.

2. Bernard Houghton, Scientific Periodicals; Their Historical Development, Characteristics, and Control, Linnet Books, Hamden, Conn., 1975.

Chapter 9

ABSTRACTING SERVICES IN MATHEMATICS

The first mathematical abstracting journal, Jahrbuch über die Fort-schritte der Mathematik, was founded in 1868. Its success was only partial because of a substantial time lag and it ceased publication in 1934. The literature of research mathematics from that time to the present day is covered by the following abstracting journals: Zentralblatt für Mathematik, founded in 1931; Mathematical Reviews, founded in 1940; and Referativnyi Zhurnal Matematika, founded in 1953. These three journals are dealt with in some detail in this chapter. Guides for locating other abstracting services that include selected mathematical literature are discussed in the final section of the chapter. Abstracting journals for the applied mathematical sciences are described in Chapters 13, 14, and 15.

I. MATHEMATICAL REVIEWS

Mathematical Reviews, published by the American Mathematical Society, is undoubtedly the most valuable abstracting journal for mathematics. For this reason, it is described first, and in detail.

Mathematical Reviews is international in scope, covering about 1,200 journals of every mathematical specialty and every language in which mathematics is written. It also includes many journals, which, by subject content are not mathematical although they contain mathematical material. An example is the inclusion of the IEEE Transactions on Aerospace and Electronic Systems. A special attempt is made to cover the mathematical activities in the USSR and the

countries within its sphere of influence. Selected books and other
nonserials as well as journals and series are reviewed. Books of a
popular nature or below the level of graduate education are not
usually included.

Selected entries from other abstracting and reviewing journals
are reprinted in Mathematical Reviews. They are: Applied Mechanics
Review, Computing Reviews, Electrical and Electronics Abstracts,
Mathematics of Computation, Operations Research, Physics Abstracts,
Referativnyi Zhurnal Matematika (Mehanika, etc.), Statistical Theory
and Methods Abstracts, and Zentralblatt für Mathematik.

In addition to the American Mathematical Society, about thirty
other national mathematically oriented societies sponsor Mathematical
Reviews; such as: London Mathematical Society, Societé Mathematique
de France, Dansk Matematisk Forening, etc. The journal was initiated
in 1940 with funds granted by the Carnegie Corporation of New York,
and over the years it has received support from the National Science
Foundation, the Rockefeller Foundation, the American Philosophical
Society, and the Alfred P. Sloan Foundation. It is presently facing
an uncertain future, due to the recent withdrawal of National Science
Foundation and private foundation funding.

A. The Reviews

The reviews are written by approximately 2,000 mathematicians,
from almost every country in the world, who are experts in the subject areas of the original papers. The reviews, which are critical
in nature, describe and evaluate the subject content of the original
papers, often reproduce factual data and the steps of proofs, and
relate the results to previous works. In reading a review, the
mathematician may gain insight into the paper beyond a mere statement of topic and results. In the "Guide for Reviewers" appears
this statement of editorial policy:

The review of a mathematical research paper should give
a concise and clear indication of its content. It is

not intended that the review be a substitute for the
paper; its primary purpose is to give the reader a
basis for deciding whether or not he should consult
the original paper....References to related work in
the literature are generally desirable....If the reviewer believes that the paper duplicates earlier work,
specific references should be offered. If the reviewer believes that the paper is in error, the errors
should be pinpointed. Criticism, if objective, docu-
mented, and expressed in good taste, is welcome [1].

Thus, Mathematical Reviews, as its name indicates, is in reality a
reviewing journal and its presentation of a paper may differ funda-
mentally from presentation of a cut-and-dry abstract in other abstracting journals.

Currently, Mathematical Reviews is published in two volumes per
year. Each volume consists of six monthly issues, with a separate
index issue for each volume. The reviews and pages in each volume
are numbered consecutively. Some idea of the exhaustiveness of this
work is given by the fact that Volume 52, July to December 1976,
contains 16,922 reviews.

When using earlier volumes of Mathematical Reviews it is necessary to be aware of some vagaries in numbering and format. Prior
to 1959 (Volume 20) the reviews were not numbered. From 1940
through 1960 (Volumes 1-21) there was one volume a year, with twelve
monthly issues and a separate index issue. In 1961 there was one
volume (Volume 22), but there were two separate parts to each monthly
issue: Part A, Pure Mathematics, and Part B, Applied Mathematics.
Each part had a separate sequence of numbers, prefaced by a letter
A or B. There is a separate index issue for this volume which covers
both Part A and Part B. In 1962, Mathematical Reviews began publishing two volumes a year, retaining the separation of the monthly
issues into two parts. However, separate indexes were issued for
Part A and Part B. So for the year 1962 there are two volumes,
Volume 23 and Volume 24; there are twelve issues for each volume,

six issues of Part A and six issues of Part B; and there are two
index issues for each volume, one for Part A and one for Part B.
Fortunately, this rather cumbersome procedure was abandoned with
Volume 25 and since 1963 there have been two volumes a year, six
issues per volume, and only one index per volume.

Reviews in the monthly issues are currently arranged under
sixty-one rather broad headings that are used in the AMS(MOS) Subject Classification. (A list of the headings is given in Table 1.)
This scheme was developed in 1970 for the Mathematical Offprint
Service, and subsequently adopted for classifying entries in Mathematical Reviews. (The scheme is described in the following section
of this chapter.) Reviews in earlier volumes are also organized
under subject groupings, but the number and terminology of the subject headings have changed over the years. For example, Volume 1
(1940) had only eleven very general headings: History, Foundations,
Algebra, Theory of Numbers, Theory of Groups, Analysis, Topology,
Geometry, Numerical and Graphical Methods, Mechanics, and Mathematical Physics. The steady increase in the number of papers reviewed
each year (there were 2,224 reviews in 1940) is one reason for revision of the classification scheme. Another factor is the literature, itself. Mathematics has its "fashions," as do other fields,
and a given topic may be an active area of research for a time and
then may be neglected temporarily. This, plus the fact that new
fields are constantly emerging, is reflected in the papers that are
published at any given time. The user should be alert to the changes
in the classification scheme of Mathematical Reviews, but he should
not be troubled by them. The cover of each monthly issue lists the
subject headings used within.

Each review is numbered and is entered under the name of the
author of the original paper or book, or the name of the first
author when there is more than one. For a book where no author is
given, entry is under the title. A complete bibliographic citation
is given for each entry, followed by a signed review. Reviews that
have been reprinted from other reviewing and abstracting journals
are identified in parentheses following the reviewer's name.

B. The Indexes

Each issue includes an alphabetical author index, giving the review number(s) for each author; a key index, which is basically an alphabetical listing of book titles with review numbers; and, a list of journal additions and changes.

The separately published index issue for each volume includes: the author and key indexes cumulated from the monthly issues; a reprinting of the AMS(MOS) Subject Classification; abbreviations of names of journals abstracted; a list of journals in translation; errata and addenda; transliteration of Cyrillic employed by Mathematical Reviews, the Library of Congress, and selected abstracting journals; and, since Volume 45 (January-June 1973), a subject index. This subject index is a classified one, arranged according to the AMS(MOS) Subject Classification. Briefly, this scheme consists of the main headings, mentioned earlier, that are used to arrange the reviews in the monthly issues, which have been further subdivided into more specific subheadings, and to which code numbers have been assigned. Each review in the subject index is listed under the classification code number(s) that have been assigned to it. The review is listed under the name of the author of the original paper and reference is to the number of the review. For example:

34   ORDINARY DIFFERENTIAL EQUATIONS

34B   Boundary Value Problems

34B15   Nonlinear boundary value problems

Gaines, R. W.	14439
Heidel, J. W.	11176
McCandless, W. L.	870
etc.	

Cumulative author indexes covering several volumes of Mathematical Reviews have been published from time to time. Those presently available are:

TABLE 1
Headings Used in Mathematical Reviews

General

History and Biography

Logic and Foundations

Set Theory

Combinatorics, Graph Theory

Order, Lattices, Ordered Algebraic Structures

General Mathematical Systems

Number Theory

Algebraic Number Theory, Field Theory and Polynomials

Commutative Rings and Algebras

Algebraic Geometry

Linear and Multilinear Algebra, Matrix Theory

Associative Rings and Algebras

Nonassociative Rings and Algebras

Category Theory, Homological Algebra

Group Theory and Generalizations

Topological Groups, Lie Groups

Functions of Real Variables

Measure and Integration

Functions of a Complex Variable

Potential Theory

Several Complex Variables and Analytic Spaces

Special Functions

Ordinary Differential Equations

Partial Differential Equations

Finite Differences and Functional Equations

Sequences, Series, Summability

Approximations and Expansions

Fourier Analysis

Abstract Harmonic Analysis

Integral Transforms, Operational Calculus

TABLE 1 (continued)

Integral Equations
Functional Analysis
Operator Theory
Calculus of Variations and Optimal Control
Geometry
Convex Sets and Geometric Inequalities
Differential Geometry
General Topology
Algebraic Topology
Manifolds and Cell Complexes
Global Analysis, Analysis on Manifolds
Probability Theory and Stochastic Processes
Statistics
Numerical Analysis
Computer Science
General Applied Mathematics
Mechanics of Particles and Systems
Elasticity, Plasticity
Fluid Mechanics, Acoustics
Optics, Electromagnetic Theory
Classical Thermodynamics, Heat Transfer
Quantum Mechanics
Statistical Physics, Structure of Matter
Relativity
Astronomy and Astrophysics
Geophysics
Economics, Operations Research, Programming, Games
Biology and Behavioral Sciences
Systems Control
Information and Communication, Circuits, Automata

20-Volume Author Index of Mathematical Reviews, 1940-1959.
This index lists every item published in the first twenty volumes
of Mathematical Reviews, with complete cross references to joint
authors, or, in the case of items without individual authors, by
editor's name or by title. Reprinted in 1966, with errata noted, it
is in two parts: Part 1, A-K, 1,092 pages; Part 2, L-Z, 1,115 pages.

Author Index of Mathematical Reviews, 1960-1964. This two-part
index is a continuation of the 20-Volume Index, covering Volumes
21-28. In addition, an account of methods of transcribing Chinese
and a table of all current transcription systems of Chinese names,
plus an errata and addenda section applicable to the 20-Volume Index,
have been included. Part 1, A-K, has 688 pages and Part 2, L-Z, has
672 pages.

Author Index of Mathematical Reviews, 1965-1972. This index is
in 4 parts and covers Volumes 29-44. The total number of pages is
3,025 and the format is the same as that of the two earlier indexes.
It is interesting to note that the first two cumulative indexes
cover 25 years in 4 volumes and contain 156,000 reviews. This latest
index covers 8 years in 4 volumes and contains 127,000 reviews.

Mathematical Reviews is currently in the process of converting
its entire data base, current and retrospective, to machine-readable
form. When this is accomplished, it is anticipated that a computer-
produced cumulative index will be published every five years. Present plans call for one last manually-produced index before then, but
the publication date is uncertain at this time.

The current data base has been computerized for the production
of the Index of Mathematical Papers which is described below.

The Index of Mathematical Papers began in 1972, not as an index
to Mathematical Reviews, but as an index of the papers received and
processed by the Mathematical Offprint Service. Since then the
scope and purpose of the Index have changed and it is appropriate to
discuss it in connection with the other cumulative indexes.

This Index covers all areas of pure mathematics; important topics
in applied mathematics and mathematical physics; and mathematical
applications in the other physical, biological, and social sciences.

Papers are listed alphabetically by author, with a complete bibliographic citation for each paper. Starting with Volume 5, issues
include all papers and books reviewed by Mathematical Reviews, with
such additional information as the location of the review, and a subject index arranged by AMS(MOS) classification numbers. The available volumes of the Index are listed below. The first dates given
in the listing are the publication dates of the Index. The dates in
parentheses are, for Volumes 1-4, the receipt dates by the Mathematical Offprint Service of the papers indexed; and, for Volumes 5 ff.,
the publication dates of the reviews of the papers and books in
Mathematical Reviews.

> Volume 1, 1972 (July-December 1970)
> Volume 2, Issue 1, 1972 (January-June 1971)
> Volume 2, Issue 2, 1972 (July-December 1971)
> Volume 3, Issue 1, 1973 (January-June 1972)
> Volume 3, Issue 2, 1973 (July-December 1972)
> Volume 4, Issue 1, 1974 (January-June 1973)
> Volume 4, Issue 2, 1974 (July-December 1973)
> Volume 5, 1975 (Index to Mathematical Reviews,
>   Volumes 45 and 46, 1973)
> Volume 6, 1976 (Index to Mathematical Reviews,
>   Volumes 47 and 48, 1974)
> Volume 7 (in preparation)

Beginning with Volume 5, the Index might best be described as
an annual index for Mathematical Reviews. Subsequently, the Index
of Mathematical Papers and Mathematical Reviews will be merged, according to the editor of Mathematical Reviews, effecting a simplification in the volume indexes of Mathematical Reviews which will be
compensated for by a timely appearance of the annual index. It
would appear, although this point is not clear, that in the future
there will be one annual index, rather than two semiannual (volume)
indexes of Mathematical Reviews [2].

By using the three basic cumulative indexes, covering 1940 to
1972, together with the Index of Mathematical Papers, starting with
Volume 5 (1973), one has access to the complete set of Mathematical
Reviews. It should be borne in mine, however, that these indexes
are basically author indexes. Searching by subject, while not impossible, is somewhat difficult. There are no alphabetical indexes of
specific subject terms, such as one finds in library catalogs or
other types of indexes. The "subject indexes" in these publications
are, in fact, classified subject indexes. A printed classified subject index might be thought of in terms of a classification scheme
used in a library to arrange books on the shelves. The books (entries,
in the case of a printed index) are organized into broad subject
groups and are then broken down into more specific sub-groups within
the broad groups, with notations attached to show their locations.
The underlying logic of the arrangement is not immediately apparent
from the notation; it is necessary to familiarize oneself with the
total classification scheme before going to the book stacks. In the
same manner, it is necessary to examine the AMS(MOS) Subject Classification (printed in each index volume) before turning to the subject
index. As noted earlier, the subject classification scheme used for
Mathematical Reviews has undergone changes over the years, so it is
wise to look at the scheme used in a given index volume before proceeding to the subject index.

II. ZENTRALBLATT FÜR MATHEMATIK UND IHRE GRENZGEBIETE

Zentralblatt für Mathematik und ihre Grenzgebiete is published
in Berlin by Springer-Verlag, with editorial offices at Deutsche
Akademie der Wissenschaften and Heidelberger Akademie der Wissen-
schaften (both in Berlin). Founded in 1931, it predates Mathematical
Reviews, and, in fact, provided the pattern for that journal. Like
Mathematical Reviews, it is international in scope, has signed ab-
stracts of the literature in various languages (chiefly German and
English), is arranged in classified order (using the AMS(MOS) Subject
Classification at the present time), and has author and subject

indexes. Over the years its frequency has varied, but currently it is published twenty-seven times a year, each issue being called a volume.

For various reasons, including the fact that it has always been published privately, it was many years before Zentralblatt recovered from the effects of World War II. In fact, publication was suspended from November 1944 until June 1948. At one time the delay between the appearance of an article and its review was between four and six years. Since 1973 Zentralblatt has been gradually changing over to the use of author abstracts in an effort to satisfy the demand for more rapid and comprehensive information about mathematical research. This procedure, according to the publisher, is intended to lead to a certain degree of differentiation and complementation between the mathematical abstracting journals. The journal has advised authors that an abstract should not be a short summary, but should comply in size and form with a normal review (although criticism is not expected), and may be written in English, German, or French. The publisher claims that abstracts are currently being published within ten weeks of receipt.

The current indexing system for Zentralblatt, which begins with Volume 101, consists of volume indexes, 10-volume indexes, and 50-volume indexes. Each volume (issue) contains an author index, a key index, a subject index, and a "Biographical Reference" section, where papers are listed under the name of the person to whom they refer. Every tenth volume is a cumulative index of the preceding nine volumes. It includes the same indexes as the volume indexes, plus a list of journals covered by Zentralblatt. The first 10-volume index is numbered Volume 110. Every fiftieth volume is a cumulation of the previous forty-nine volumes. The first 50-volume index is Volume 300. (For unexplained technical reasons Volume 250 only includes data from Volumes 221-249.)

A retrospective indexing of Volumes 1-100 was undertaken in 1974 and to date, indexes have been published for the following years:

1931-1940. This is a two-part index, numbered Volume 62I and Volume 62II, for the Volumes 1-25. It is an index of authors only.

1941-1950. This is a cumulative author index to Volumes 26-41 and 60-61, and is numbered Volume 63I and Volume 63II.

1951-1956. Annual indexes were published for these years. The numbering of the index volumes is as follows: Volume 54 (for 1951); Volume 69 (1952/53); Volume 59 (1954); and, Volume 76 (1955/56).

III. REFERATIVNYĬ ŽHURNAL MATEMATIKA

Referativnyĭ Žhurnal has been issued monthly since 1953 by the Vsesoyuznogo Instituta Nauchnoi i Tekhnicheskoi Informatsii (All-Union Institute of Scientific and Technical Information), Moscow. It is a major abstracting journal for the world's literature in most branches of science and technology.

Referativnyĭ Žhurnal consists of several series, designated svodnyi tom (joint volume), each devoted to an individual branch of science or technology. The number and titles of the series have varied over the years. A series, or joint volume, is made up of chapters that are issued as separate periodicals with their own titles, and designated vypusk RZh (section of the Referativnyĭ Žhurnal). The mathematical series is titled Referativnyĭ Žhurnal Matematika and consists of three chapters: (1) General, Mathematical logic, Theory of numbers, Algebra, Topology, Geometry; (2) Mathematical analysis; and, (3) Numerical analysis, Probability theory and mathematical statistics, Cybernetics.

Referativnyĭ Žhurnal Matematika is practically identical in format, style, and purpose to Mathematical Reviews, except that all the reviews are written in Russian. The reviewing staff comprises almost the entire professional community of the USSR, together with a relatively small number of Eastern Europeans. The coverage is about the same as that of Mathematical Reviews, except that a large number of pedagogical articles are listed in Referativnyĭ Žhurnal Matematika. Books are usually listed in Referativnyĭ Žhurnal Matematika but reviewed elsewhere.

Bibliographic information in Referativnyĭ Žhurnal Matematika is listed in the original language following the Russian entry. There is a monthly author and subject index.

During the 1950s, Mathematical Reviews established extensive reciprocity arrangements with both Zentralblatt für Mathematik und ihre Grenzgebiete and Referativnyĭ Zhurnal Matematika, by which Mathematical Reviews may reprint reviews appearing in those journals, and vice versa. The purpose was to prevent excessive delay in the appearance of the review of a paper, since all three journals reflect almost identically the needs of the working mathematician. In no sense can the three abstracting journals be thought of as competitive, however, for different reviews of the same paper can give perspective to the research involved.

Since the founding of Mathematical Reviews, several new abstracting or reviewing journals have been established in certain peripheral areas of coverage of Mathematical Reviews; for example, in the fields of computer science and operations research. Mathematical Reviews encouraged these new journals by making available information concerning materials in their areas and supplying lists of potential reviewers. Reciprocity arrangements were made with them, too, whereby reviews from one journal might be reprinted in another. The major abstracting journals with which Mathematical Reviews has reciprocity arrangements of some sort have been mentioned on page 68 [3].

IV. OTHER ABSTRACTING SERVICES

The purely mathematical abstracting journals may be supplemented, especially for subjects of an applied or interdisciplinary nature, by abstracting journals that are devoted primarily to other subjects. Applied Mechanics Reviews and Computer and Control Abstracts, for example, are two journals that abstract selected mathematical literature. There are two bibliographies, which, although somewhat out-of-date, are still useful for locating additional abstracting services.

A Guide to the World's Abstracting and Indexing Services in Science and Technology, National Federation of Science Abstracting and Indexing Services, Washington, D.C., 1963, is an alphabetical title listing of 1,855 services in all fields of science and technology. Complete bibliographic data is given for each title: date the publication began, frequency, average number of abstracts per year, subject coverage, method of arrangement, and price. The main sequence is supplemented by a subject index, a country index, and an index of services arranged by the Universal Decimal Classification.

Abstracting Services, published by the International Federation for Documentation, The Hague, is in two volumes, and covers all subject fields. Volume 1, Science, Technology, Medicine, Agriculture, 2nd edition, 1969, lists approximately 13,000 abstracting services in those fields. Arrangement is alphabetical by title and is supplemented by a country index, a subject index, and a classified listing, arranged by the Universal Decimal Classification. Information given for each service includes the number of abstracts published each year, length of abstracts, and the number of journals monitored by the service.

To locate new services that have appeared since the above bibliographies were published Ulrich's International Periodicals Directory is a useful source. As mentioned earlier, Ulrich is organized alphabetically by subject and the first subject in the list is "Abstracting and Indexing Services." Under this heading abstracting journals are listed alphabetically, together with information about date of origin, frequency, publisher, cost, and a brief description of coverage.

REFERENCES

1. "Mathematical Reviews: Guide for Reviewers," unpublished, 4 pp.

2. R. G. Bartle, Executive Editor, Mathematical Reviews, personal correspondence, January 19, 1977.

3. A. J. Lohwater, "Mathematical Literature," Library Trends, 15, 852-867 (1967).

Chapter 10

BOOKS

I. GENERAL

It is often said, somewhat scornfully, that in science and technology
books contain only second-hand knowledge. While it is true that the
journal is the better medium for communication of current research
results, the role of the book should not be underestimated. Books
serve to introduce the layman to the general field of the science,
or the scientist to a field outside his own specialty; books can
explain a new theory in light of already known facts; and books help
to coordinate and systematize knowledge. E. J. Crane's remarks about
chemical books apply equally well to mathematical books: "historical
works record the development of a subject, popular works initiate
the public into its mysteries and elicit interest, and treatises
give the reader the benefit of the long experience or combined re-
searches of many workers" [1].

The forms that books take most often are the treatise, the monograph, and the textbook. Traditionally, the main difference between
the treatise and the monograph is that the treatise attempts to cover
the whole of its subject field and the monograph deals with a nar-
rowly-defined single topic. The textbook is a teaching instrument
whose primary aim is to develop an understanding of its subject,
rather than impart exhaustive information about it [2]. There is
danger in trying to make too fine a distinction among these forms,
however. One can always find examples where the forms overlap, and
often the titles of books can be misleading. More importantly,

different people may use the same book for different purposes, depending upon the type of information required. This is particularly
true in mathematics, where a given book may serve as a monograph for
the nonspecialist who wishes to familiarize himself with a new subject, a textbook for the student, or even as a reference book for the
specialist who needs to check a specific piece of information.

For convenience in discussing the various types of mathematical
books, they have been grouped into three categories: research, exposition, and study. This reflects the primary purpose of the books
in each group, but not the sole use to which they may be put.

As was the case with journals, mathematical books are published
by societies, universities and academies, and commercial publishers.
In the area of research mathematics the American Mathematical Society
plays such an important publishing role that its publications are
discussed separately and before those of university and commercial
publishers.

II. RESEARCH: AMERICAN MATHEMATICAL SOCIETY PUBLICATIONS

The American Mathematical Society publishes several series of
monographs, proceedings of symposia, translation series, and review
volumes. These may be characterized as advanced research-level publications and are representative of the types of publications produced by professional societies in other countries, as well.

A. Monograph Series

The Society's three monograph series are Colloquium Publications,
Mathematical Surveys, and Memoirs of the American Mathematical Society.

Colloquium Publications is the oldest of the monograph series,
having started in 1905. It contains syntheses of recent and older
mathematical work prepared by outstanding research mathematicians.
The usefulness of this series is evidenced by the fact that every
title has been revised and/or reprinted at least once. An illustra-
tion of this is Foundations of Algebraic Geometry by Andre Weil which

was first published in 1946, revised and enlarged in 1962, and reprinted in 1975. Some early titles have been allowed to go out of
print, but they are still available from Xerox University Microfilms.
In addition to the texts themselves, the bibliographies included in
these books are of great value. In conducting a retrospective or
exhaustive search, one can begin with the references in the monograph and then go forward in time, from the publication date of the
monograph, adding more recent references.

Mathematical Surveys is a series that started in 1943 to meet
the need for expositions in fields of current interest in research.
Each of the books gives a survey of the subject and an introduction
to its recent developments and unsolved problems. Volumes are published irregularly and currently there are 14 titles in the series.

Memoirs of the American Mathematical Society, which began in
1950, contains research tracts of the same general character as the
papers published in the journal Transactions of the American Mathematical Society. As issue of the Memoirs contains either a single
monograph or a group of cognate papers. All of the issues in this
series (which numbered 153 at the beginning of 1975) are in print.
Many of the earlier numbers have been revised and/or reprinted.
Since 1975 Memoirs has been available as a journal subscription and
since 1977 it has been issued bimonthly.

B. Proceedings of Symposia

Proceedings of Symposia in Pure Mathematics is a series of
books containing papers on various branches of pure mathematics.
These papers were presented at symposia and summer research insti-
tutes held under the auspices of the American Mathematical Society
and other organizations, since 1959. Some of the volumes are part
of a subseries published for the Association of Symbolic Logic by
the Society.

The American Mathematical Society publishes several other series
of proceedings of symposia in applied mathematics. These are discussed in Chapter 13.

C. Translation Series

American Mathematical Society translations are published in the form of books, but the original sources may be in the form of books, papers, or proceedings.

Translations of Mathematical Monographs, as the title indicates, is a series of translations of separately published books, chiefly from Soviet sources. The series contains both advanced mathematical research works and expository works.

American Mathematical Society Translations - Series 1, consists of 105 separately published research papers that were selected from the Russian and other Eastern European and Oriental languages. The papers were later grouped by subject and published in 11 volumes. For example: Volume 1, Algebra; Volume 9, Lie Groups.

American Mathematical Society Translations - Series 2, is a sequel to the earlier Series 1. It is made up of research papers that in the original languages were published separately. But since this is an ongoing series, numbering well over 100 volumes to date, the papers are not grouped by broad subjects, as is the case in Series 1. Some volumes may be devoted to a single subject, but others may include two or more topics. An example of the former type is Volume 22, Nine Papers on Analysis. An example of the latter type is Volume 33, Eleven Papers on Differential Equations and Two on Information Theory.

The Index to Translations Selected by the American Mathematical Society requires some explanation, since its title is rather misleading. Volume 1 (1966) covers American Mathematical Society Translations - Series 1, American Mathematical Society Translations - Series 2, Volumes 1-50, and Selected Translations in Mathematical Statistics and Probability, Volumes 1-5. (This last title is discussed in Chapter 14.) Volume 2 of the Index (1973) covers American Mathematical Society Translations - Series 2, Volumes 51-100, and Selected Translations in Mathematical Statistics and Probability, Volumes 6-13.

Both volumes of the Index contain author and subject indexes with complete bibliographic information for each original paper

published in these series, as well as the American Mathematical
Society series title, volume, and pages.

Proceedings of the Steklov Institute of Mathematics is a cover-
to-cover translation of the proceedings of the Steklov Institute of
Mathematics in the Academy of Sciences of the USSR, beginning with
the fifteen issues dated 1965 and 1966 in the original. Each issue
contains either one book-length article or a collection of articles
on one topic. The topics contained in these volumes are all in the
forefront of contemporary mathematical research.

Transactions of the Moscow Mathematical Society is another cover-
to-cover translation. This is a translation of the Soviet journal
Trudy Moskovskogo Matematiceskogo Obscestva, which contains the results of original research in pure mathematics by many of the best
mathematicians in the Soviet Union, as well as by some non-Soviet
mathematicians. Each volume contains about twelve papers, approximately 400 pages. The translations begin with Volume 12 (1963) of
the Soviet journal.

D. Review Volumes

Another type of publication, one that is especially useful for
retrospective or exhaustive searching, is the review volume. This is
a compilation of reviews devoted to a particular topic, or group
of related topics, that appeared in Mathematical Reviews within a
given time period. To date, four such works have been published,
with more planned for the future.

Reviews of Papers in Algebraic and Differential Topology, Topological Groups, and Homological Algebra (1969), edited by N. E.
Steenrod, contains 6,000 reviews which appeared in Mathematical Reviews between 1940 and 1967. These reviews, with an author index,
are listed under 290 subject headings.

Reviews on Infinite Groups (1974), edited by Gilbert Baumslag
contains 4,563 reviews, which appeared between 1940 and 1970. The
reviews are arranged under 264 subject headings in this 2-volume
set.

Reviews on Finite Groups (1974), edited by Daniel Gorenstein, contains 3,052 reviews in one volume. These reviews, which are arranged under 223 subject headings, appeared between 1940 and 1970.

Reviews in Number Theory (1974), edited by W. J. LeVeque, contains 14,426 reviews which appeared between 1940 and 1972. The reviews are arranged under 282 subject headings. This is a 6-volume set, with the last volume containing the author index.

Of special interest are the forward citations given at the end of many of the reviews in these volumes. Such citations take note of any references made to the paper at a later date by a review or reviews in the collection.

The advantage of a review volume of this type is that it brings together in one location material that previously had to be searched in several volumes. One must bear in mind, however, that only reviews that appeared in Mathematical Reviews are included.

III. RESEARCH: OTHER PUBLISHERS

There has been an increasing tendency on the part of publishers in the field of mathematics to establish monograph series under the guidance of a competent editor. It is not possible to discuss all of these series in this guide; the titles described below are some of the important series and are representative of this type of publication.

Ergebnisse der Mathematik und ihrer Grenzgebiete is issued by a commercial publisher, Springer-Verlag of Berlin. Each volume in this monograph series may be described as a state-of-the-art treatment of an individual topic. The treatment of a topic is exhaustive, bringing together and relating results that previously were scattered throughout the literature, primarily the journal literature. Of immeasurable value to the researcher is the extensive bibliography, arranged chronologically, which appears at the end of each volume. German and English are the languages in which these books are published.

Lecture Notes in Mathematics is another Springer publication.
This series attempts to report quickly new developments in mathematical research and teaching. The type of material in this series
includes preliminary drafts of original papers, lectures on a new
field, or presentations of new perspectives on a classical field,
and papers from seminars. Each volume is devoted to a single topic
and the language may be German, French, or English. In 1976 Springer
published an index for Volumes 1-500 of Lecture Notes, including
both author/editor and subject indexes.

Cambridge Tracts in Mathematics and Mathematical Physics is a
monograph series published by Cambridge University Press since 1905.
The purpose of this series is to provide introductions to modern
topics in mathematics. While the books are not intended for specialists, they cannot be considered elementary in the conventional sense.

IV. EXPOSITION

An expository book in mathematics, as noted earlier, can by no
means be considered a "popular" work. The term is used to describe
a work that introduces, synthesizes, or traces the development of a
topic. In this sense it can serve research mathematicians who are
not familiar with the topic, students and teachers of mathematics,
workers in other fields of science, or members of the general public
who have sufficient knowledge of mathematics.

The Mathematical Association of America publishes two important series of expository books: the Carus Mathematical Monographs
and the MAA Studies in Mathematics.

The Carus Mathematical Monographs is a series of books intended
to make topics in pure and applied mathematics accessible to students
and teachers of mathematics. The scope of this series also includes
historical and biographical monographs. Each book is written by an
authority in the field and contains an extensive bibliography.

The MAA Studies in Mathematics is a series devoted to recent
developments in mathematics, both pure and applied. Each volume is

a collection of papers on a given topic, by various experts in the
field. Again, the presentation is at the collegiate and graduate
level, but the books have appeal for a wider audience.

Two other Mathematical Association of America series, the Dol-
ciani Mathematical Expositions, which are collections of essays, and
the Selected Papers series, which are volumes of reprints of papers
that appeared in The American Mathematical Monthly and Mathematics
Magazine, are also designed to appeal to a broad audience.

The CBMS Regional Conference Series in Mathematics, published
by the American Mathematical Society, is another publication of this
type. Each monograph in the series is an expository paper developed
by the author from lectures which he or she, as principal speaker,
presented at one of the regional conferences that the Conference
Board of the Mathematical Sciences has sponsored since 1970.

The Mathematical Sciences: A Collection of Essays, edited by
the Committee on Support of Research in the Mathematical Sciences of
the National Research Council, was published in 1969 for the National
Academy of Sciences, by M.I.T. Press. This book is a companion volume to The Mathematical Sciences: A Report, prepared by the same
Committee and published the previous year. The Report, the result
of a study to assess the status and projected future needs of the
mathematical sciences, is now out-of-date, many of the projections
having proved inaccurate. The Essays, written by distinguished mathematicians on various topics in pure and applied mathematics, was intended to describe what constitutes contemporary mathematical research. Addressed to the nonmathematician scientist and the scien-
tifically-oriented layman, this book is nonetheless substantive
enough to be of interest to the mathematician and student, as well.

V. STUDY

This category has deliberately been labeled "study," rather
than "textbooks." The titles discussed here are intended primarily
to serve as texts for students of mathematics, but, as noted earlier,

such books may also be useful to nonspecialists and scientific workers in other fields.

Most books of this type are produced by commercial publishers, and generally in the form of book series. Many may be considered monographs, but the majority are intended to be used as texts. Also, a large number of these series include books in both pure and applied mathematics. A case in point is Pure and Applied Mathematics: A Series of Monographs and Textbooks, published by Academic Press. Some other titles of series are:

Addison-Wesley's Advanced Book Program, which encompasses all the sciences, includes graduate-level texts in mathematics.

North-Holland Publishing Company and its American affiliate American Elsevier Publishing Company publish several series of graduate-level texts. Some of their titles are: Advanced Studies in Pure Mathematics; Bibliotheca Mathematica; North-Holland Mathematical Library; North-Holland Mathematics Studies; and, North-Holland Texts in Advanced Mathematics.

Springer-Verlag publishes two textbook series: Graduate Texts in Mathematics and Undergraduate Texts in Mathematics.

These are only a few of the many commercial firms that publish mathematical books. Additional names and addresses of publishers may be found in directories such as the American Book Trade Directory, published biennially by Bowker, New York. Information on obtaining additional book titles is discussed in the next chapter.

REFERENCES

1. E. J. Crane, A Guide to the Literature of Chemistry, 2nd edition, Wiley, New York, 1957.

2. Denis Grogan, Science and Technology: An Introduction to the Literature, 2nd edition, Linnet Books, Hamden, Conn., 1973.

Chapter 11

ACCESS TO BOOKS: BIBLIOGRAPHIES AND REVIEWS

Finding the books to meet one's needs at any given time involves the
processes of identification (discovering what books are available)
and evaluation (determining whether the books satisfy one's needs).
The tools used for identification (bibliographies) and evaluation
(reviews) are the subjects of this chapter.

I. BIBLIOGRAPHIES

A bibliography is a list of references to literature on a particular place, person, or subject. A bibliography may be "mixed";
that is, it may include both books and nonbook materials. However,
it is customary to use the term bibliography for lists of books and
the term index for lists of nonbook materials.

Bibliographies indicate the existence of books without neces-
sarily indicating where they may be obtained. In this way they
differ from library catalogs, which list the holdings of a particular
library or group of libraries, and publishers' and booksellers'
catalogs, which tell where the books may be purchased.

Bibliographies may be very simple lists, giving only the author,
title, publisher, date and place of publication. Or they may be
very complex devices that give a physical description of the book,
tell whether a list of references is included, provide an annotation, and so forth.

Entries in bibliographies are arranged in one of the following
ways: author; chronologically; alphabetical subject; classified

subject; or, dictionary. A few minutes spent in ascertaining the
arrangement of a given bibliography will greatly expedite its use.

Another factor to consider when using a bibliography is the
date limits of the material listed. This information will be indicated somewhere in the work, usually in the preface.

Bibliographies may be issued as complete works, as serials, or
they may appear as parts of other journals and books. Theodore
Besterman's World Bibliography of Bibliographies, 3rd edition, So-
cietas Bibliographica, Geneva, 1955-56, 4 volumes, reprinted Scare-
crow Press, New York, 1960, 2 volumes, is a bibliography of separately published bibliographies. More than 80,000 bibliographies
are listed, arranged alphabetically by subject, with an author index, and an indication of the number of items included in each bibliography. Bibliographic Index: A Subject List of Bibliographies
in English and Foreign Languages, Wilson, New York, 1937-, is an
alphabetical subject index to bibliographies that appear in about
1,500 journals. It is published three times a year, with the third
issue a cumulation of all three issues.

A work that is, strictly speaking, not a bibliography is nonetheless useful for its coverage of the early mathematical literature.
This is J. C. Poggendorff, Biographisch-literarisches Handwörterbuch der exakten Naturwissenschaften, Barth, Leipzig, 1864-1940, 6
volumes in 10, reprinted, Edwards Bros., Ann Arbor, Mich., 1945.
Poggendorff is a bio-bibliographic work covering all the important
mathematicians, astronomers, physicists, chemists, and other scientists of all countries, for all times up to and including 1931. The
work is arranged alphabetically by name of scientist, with a brief
biographical sketch followed by a detailed bibliography of his
writings.

Volume 3 of Les Sources du Travail Bibliographique, by L. N.
Malclès, Droz, Geneva, 1950-58, covers the natural sciences. Mathematical sciences are dealt with in detail in this volume under the
headings "Partie Generale" and "Partie Speciale." Although this is
a guide to reference materials of all kinds, more than half of the
titles are those not commonly classified as reference works. For

example, included are histories, journals, and textbooks, mainly
from European and American publishers.

Part II of N. G. Parke, Guide to the Literature of Mathematics
and Physics, Including Related Works on Engineering Science, 2nd
edition, Dover, New York, 1958, is a selected bibliography of about
5,000 entries grouped under approximately 120 subject headings.
Under each subject heading is a brief description of the subject,
followed by the entries arranged alphabetically by author. There is
an author index and a subject index.

E. M. Dick, Current Information Sources in Mathematics, Libraries Unlimited, Littleton, Colo., 1973, continues where Parke left
off. This book is a selective bibliography of books published or reprinted in English or in English translation from 1960 to mid-1972.
The emphasis is on monographic material, primarily for students and
teachers. The 1,600 entries are arranged in 37 chapters: 33 devoted
to branches of mathematics; and 1 each for periodicals, guides and
directories, professional organizations, and publishers.

In 1961, the American Bibliographic Service (Darien, Conn.) instituted a new service, Quarterly Checklist of Mathematica: An International Index of Current Books, Monographs, Brochures and Separates.
Unfortunately, publication ceased with Volume 10 in 1970, but the
volumes that were issued from 1961 to 1970 may be used to supplement
the Dick book for the literature published during that period.

The best source for books and papers being published currently
is the biweekly journal Current Mathematical Publications. Issued
by the American Mathematical Society since 1975, this is a classified
subject bibliography of materials received by Mathematical Reviews
for review in that journal; therefore, it is not comprehensive.
Entries are arranged by the AMS(MOS) Subject Classification, and
there is an author index and a key index for each issue. This journal was formed by the merger of "American Mathematical Society. New
Publications," which was a department of the Notices of the American
Mathematical Society, and Contents of Contemporary Mathematical
Journals, a current awareness journal that the Society had published
since 1969. Contents reprinted the tables of contents of about 240

mathematical journals. In 1972 the title was changed to Contents of
Contemporary Mathematical Journals and New Publications when the
scope of the journal was enlarged to include books and other separately published materials. In March of 1974 the format of Contents
was completely revised. It no longer reproduced title pages of journals, but adopted the format described above. In January 1975 the
present title was adopted.

Special mathematical bibliographies exist for individual branches
of mathematics. A recent example is K. O. May, Bibliography and Research Manual of the History of Mathematics, University of Toronto
Press, 1973. This is a classified, annotated bibliography of the
literature on the history of mathematics. It is preceded by a brief
manual on the methodology of research, and it is supplemented by
author and subject indexes.

Karl Zeller and Wolfgang Beekmann, Theorie der Limitierungsver-
fahren, 2nd edition, Springer-Verlag, Berlin, 1970, is the standard
bibliography in Series and Summability. The first section of this
book is an outline of the subject, with references incorporated
within the text. The second section is the bibliography proper,
arranged chronologically (covering the years 1880-1968), and then
alphabetically by author. This is a volume in the series Ergebnisse
der Mathematik und ihrer Grenzgebiete and most of the other volumes
in the series follow this same pattern.

Specialized journals often contain bibliographies; for example,
Bibliography of Symbolic Logic, compiled by Alonzo Church, appeared
as Volume 1, Number 4, and Volume 3, Number 4 of the Journal of
Symbolic Logic. It is also available as a separate publication from
the American Mathematical Society. "A Bibliography on Fuzzy Sets,"
compiled by J. L. F. DeKerf, is found on pages 205 to 212 of Volume
1, Number 3 (1975) of the Journal of Computational and Applied Mathematics.

Other bibliographies of the literature of the applied mathematical sciences are discussed in Chapters 13, 14, and 15.

As the reader has no doubt inferred from the above remarks,
there are no separately published bibliographies that are devoted

solely to mathematics or that cover the mathematical literature comprehensively. One must extract the mathematical literature from the
more general bibliographies and use many special-purpose mathematical
bibliographies. Additional sources are the general international
bibliographies, the catalogs of the great national libraries, and the
national and trade bibliographies. Space does not permit a discussion
of these types of bibliographies and for information about them the
reader is referred to the guides to reference books listed at the
end of this chapter [1], [2].

In actual practice, a mathematician depends most heavily on the
bibliographies that are appended to monographs. Beginning with a
book on a given topic, he will follow the references listed there,
proceeding in the fashion described in the section on exhaustive
searching in Chapter 5. Often, having obtained the classification
number of a book known to him, the mathematician will go to the library shelves and look at the other books classified by that number.
Browsing in this fashion will reveal some additional titles, but it
has obvious limitations.

II. REVIEWS

Some of the bibliographies already described are inherently
evaluative. That is, by virtue of their arrangement according to
user requirements, subject scope, time coverage, annotations, or by
the prestige of their compiler or sponsoring body, they can assist
in making a preliminary selection of books. However, a more useful
aid to selection, short of examining the actual volume, is the book
review.

Many mathematical journals include book review sections. As a
matter of course, the mathematician will read the book reviews in
the journals which he regularly reads. The Bulletin of the American
Mathematical Society contains lengthy reviews of advanced books of
interest to the research mathematician. The American Mathematical
Monthly reviews college-level books, which are of interest to the
teacher of mathematics. The Monthly also contains a section entitled

"Telegraphic Reviews." The telegraphic reviews, which are designed
to give prompt notice of new books, are very brief, with only enough
information to assist the reader in deciding whether to order a copy.
The factual data is coded: T = textbook; P = professional reading;
S = supplementary reading; L = undergraduate library purchase;
13 to 18 = freshman to second year graduate level usage; 1 to 4 =
appropriate time in semesters to cover text. Asterisks (*) or question marks (?) denote special positive or negative emphasis, respectively.

Mathematical Reviews, as noted earlier, reviews books as well
as journal articles. Book titles are indicated by an asterisk (*),
making them stand out from among the journal entries. The Mathematical Reviews Cumulative Author Indexes (discussed in Chapter 9)
and the Review Volumes (discussed in Chapter 10) are used to locate
reviews of older books.

Two publications that are primarily book selection tools compiled by and for librarians are also useful to the mathematician in
keeping abreast of current books. The ASLIB Book List is an annotated list of recommended scientific and technical books. Published
monthly, since 1935, by the Association of Special Libraries (London),
the entries are arranged by the Universal Decimal Classification and
graded according to mathematical level (general reader, student textbooks, advanced, etc.). The Technical Book Review Index is published
monthly by the Special Libraries Association (New York). Under each
book listed, extracts from published reviews are given. As its title
states, this publication serves as an index to book reviews which
have appeared in over 2,500 journals. The Index began publication
in 1935 also.

Since 1965, the American Association for the Advancement of
Science has published Science Books: A Quarterly Review, a journal
consisting entirely of reviews. In 1975 the title was changed to
Science Books and Films, to reflect its expanded coverage. This
journal includes selected mathematical books and is another source
of information about current publications.

REFERENCES

1. E. P. Sheehy, ed., Guide to Reference Books, 9th edition, Amer-
    ican Library Association, Chicago, 1976.

2. A. J. Walford, ed., Guide to Reference Material, Volume 1, Sci-
    ence and Technology, 3rd edition, The Library Association,
    London, 1973.

Chapter 12

REFERENCE BOOKS

I. GENERAL

It may seem somewhat artificial to distinguish between reference and
nonreference books. Any work, of course, can be used for reference;
that is, can be "looked up" for a specific piece of information.
This is particularly true in mathematics, where there are relatively
few reference books, and where monographs and textbooks are often
used in this way. However, the characteristic of a true reference
book is that it is designed to organize the primary literature according to some definite plan in order to facilitate quick consultation. Because reference books are used in this way and not read
from cover-to-cover, it is customary for libraries to shelve them in
a separate section and not permit them to circulate.

The types of reference materials that are available in mathematics are: guides and manuals; encyclopedias; dictionaries; handbooks and tables; and, historical and biographical works. Indexes,
abstracts, and bibliographies are other types that have already been
discussed in earlier chapters.

II. GUIDES AND MANUALS

Guides and manuals are usually aimed at a wide public: students,
teachers, research workers, practicing mathematiciams, and librarians.
A guide is probably the best place to begin when studying the literature of any subject. A guide differs from a subject bibliography,

which is basically a list of references to the literature, by including discussions of the functions and uses of the various types
of literature [1].

An example of this type of reference book is J. E. Pemberton,
How to Find Out in Mathematics, 2nd edition, Pergamon Press, Oxford,
1969. Supplementing the chapters on the various types of mathematical literature (reference books, journals, etc.) are discussions
of careers in mathematics, mathematical societies, evaluation and
acquisition of books, and mathematics and the government.

Another useful work of this type is N. G. Parke, Guide to the
Literature of Mathematics and Physics, Including Related Works on
Engineering Science, 2nd edition, Dover, New York, 1958. Part I of
this book is comprised of chapters on the principles of reading and
study, searching the literature, types of materials, and library
usage. Part II is the bibliographic part of the guide and has been
described in the previous chapter.

III. ENCYCLOPEDIAS

Encyclopedias have often been criticized by the specialist as
being superficial and out-of-date. It must be remembered, however,
that the role of the encyclopedia is not to say the last word on a
subject, but to offer a starting point for a closer investigation
of a topic. The articles in reputable encyclopedias are written by
experts in their fields and the bibliographies appended to the articles are, in themselves, of value [1].

For a general introduction and for information about major topics and well-known people in mathematics, the general multi-volume
encyclopedias can be very useful. The 9th, 11th, and 14th editions
of the Encyclopaedia Britannica are particularly good. In editions
since the 14th, the previous monograph-length articles have been
replaced by shorter articles. In all editions, however, the articles
have been written by the leading specialists of the day. As a means
of keeping up-to-date, the Britannica issues yearbooks containing

authoritative brief accounts of main events in all fields (including mathematics) for each year.

There are also multi-volumed encyclopedias devoted to science and technology. These generally offer more detailed information than the corresponding sections of general encyclopedias, and their bibliographies may be more extensive. An example of this type of encyclopedia is the McGraw-Hill Encyclopedia of Science and Technology, 2nd edition, New York, 1966. The basic 15 volumes are continuously revised and since 1962, have been supplemented by yearbooks.

An encyclopedia devoted solely to mathematics is the monumental Encyklopädie der Mathematischen Wissenschaften mit Einschluss ihrer Anwendungen, 6 volumes in 23, Teubner, Leipzig, 1898-1935. Despite its age it remains a valuable source of reference, for its period of publication spans one of the most fruitful periods of mathematical research. Noted for its comprehensive treatment and well-documented, scholarly articles, it is aimed at the specialist. In 1950 a second edition of the Encyklopädie was begun, but it is only slowly replacing the original edition.

As recently as 1976 there was no English language encyclopedia of mathematics. In December of that year, the Addison-Wesley Publishing Company issued the first two volumes of its Encyclopedia of Mathematics and Its Applications, under the general editorship of Gian-Carlo Rota. The purpose of the Encyclopedia is to present in each volume a thorough, comprehensive account of one area of mathematics of current interest. The style of exposition differs from that of the usual research paper in that mathematics is considered as a body of facts to be explained, not only by proof, but by examples and applications. The material is intended to be accessible to nonspecialists, especially non-mathematicians. Each volume includes the usual references, a guide to the literature of the subject, author and subject indexes. Results whose proofs could not be included because of space limitations are given as notes, examples, or exercises. The Encyclopedia is not arranged alphabetically, but is subdivided into sections, each comprising a broad area of mathe-

matics, such as differential equations or probability. Volumes will
be numbered in order of publication, but the exact number of volumes
in each section and the date of completion have not been determined.
It is the aim of the Encyclopedia eventually to cover all of the
branches of present-day mathematics. It is the hope of the publisher
that the volumes can be used as texts and guides for research, as
well as for reference. The volumes published to date are:

> Volume 1. Integral Geometry and Geometric Prob-
> ability, by L. A. Santalo. Section: Probability.

> Volume 2. The Theory of Partitions, by G. E. Andrews.
> Section: Number Theory.

In 1977 the M.I.T. Press published an English translation, by
Kenneth O. May, of Encyclopedic Dictionary of Mathematics, by the
Mathematical Society of Japan, edited by Shokichi Iyanaga and
Yukiyosi Kawada. This two-volume work does not merely define terms,
as its title would indicate, but describes what is generally recognized as all areas of advanced mathematics and, in most of these
areas, the status of current research. Each article is prefaced
with general remarks or a history, subdivided into appropriate,
cross-referenced sections, and followed by a list of relevant research
papers and books. Technical terms are set in boldface type and defined at the time they are initially introduced. There is a subject
index and a name index, citing each name appearing in the text or
references, as well as the name, or names, identified with an entry.
This is not a layman's encyclopedia, but rather, it is designed for
use by students and professional scientists.

IV. DICTIONARIES

A dictionary of the terminology in current use is a valuable
reference tool for the mathematician. This type of book is needed
because mathematics has become so specialized that the terminology

in one area of research mathematics is today virtually incomprehen-
sible to a worker in another area [2].

Some mathematical dictionaries are encyclopedic in nature, presenting condensations of mathematical concepts, not mere word definitions. An outstanding English language dictionary of this sort is
R. C. James, Mathematics Dictionary, 4th edition, Van Nostrand,
Princeton, N.J., 1976. Appendixes to this work contain a selection
of tables, a list of mathematical symbols, and there are indexes in
French, German, Russian, and Spanish.

A more comprehensive dictionary is the two-volume Mathematisches
Wörterbuch, mit Einbeziehung der theoretischen Physik, edited by
J. Nass and H. L. Schmid, Pergamon Press, Oxford, 1961. The treatment of each topic is thorough, sometimes including bibliographies,
and there are brief biographies of about 400 deceased mathematicians.
Translations of terms into English, French, and Russian is planned
for the next edition.

Probably the most comprehensive dictionary is Iwanami Sūgaku
Jiten (Iwanami Mathematical Dictionary), 2nd edition, Iwanami Shoten,
Tokyo, 1968. This work defines virtually every term used in mathematics and gives references for further reading. The terms are
listed alphabetically in romanized form of Japanese, followed by
native Japanese form, with English, French, German, and Russian
equivalents. The definitions are in Japanese. There is an English,
French, German and Russian index of terms, and an English index of
names. Although the Japanese text presents difficulties for many
readers, the multi-language index permits users to find the references
to a given topic within the text.

Another kind of dictionary is the bilingual or multilingual one.
In this type, the definitions, not just the indexes, are in two or
more languages. A dictionary of this kind is useful when reading in
a foreign language and the need arises to know the English equivalent
of a foreign word or phrase, rather than the mathematical concepts
involved.

The Russian-English Dictionary of the Mathematical Sciences, by
by A. J. Lohwater and S. H. Gould, American Mathematical Society,

Providence, R.I., 1961, is an example of this type. The English-
Russian Dictionary of Mathematical Terms, Izdatel'stvo Inostrannoĭ
Literatury, Moscow, 1962, was compiled as a companion volume to the
Lohwater and Gould work, in a joint project of the National Academy
of Sciences of the United States, the Academy of Sciences of the
U.S.S.R., and the American Mathematical Society.

Most libraries and most individuals acquire bilingual dictionaries in the more widely used languages: French, German, Italian,
Russian, and Spanish. When it is necessary to identify works in the
less common languages, the following compilations, although somewhat
out-of-date, are useful:

U.S. Library of Congress, Foreign Language-English Dictionaries,
Volume 1: Special Subject Dictionaries, With Emphasis on Science
and Technology, 1955. The entries in this bibliography are arranged
alphabetically by subject. Within each subject group, bilingual
dictionaries are listed first, then multilingual dictionaries.

UNESCO, Bibliography of Interlingual Scientific and Technical
Dictionaries, 4th edition, Paris, 1961, and Supplement, 1965. Entries in this bibliography are arranged according to the Universal
Decimal Classification. There are English, French, and Spanish indexes of authors and subjects.

Dictionaries of terminology in fields of applied mathematics
represent still another type and are discussed in Chapters 13, 14,
and 15.

V. HANDBOOKS AND TABLES

Handbooks are compilations of miscellaneous information in
handy form, designed for day-to-day consultation on the job, when a
straightforward factual problem arises. A handbook is often the
quickest source for a formula or tabular value. It is probably the
most frequently consulted reference tool of the working scientist
and technologist; it is used less frequently by the research mathematician. For example, the preface to Mathematical Handbook for
Scientists and Engineers, by G. A. Korn and T. M. Korn, 2nd edition,

McGraw-Hill, New York, 1968, states that this comprehensive collection of definitions, theorems, and formulas is designed specially
for engineers, scientists, and others whose work involves mathematics and its methodology. A more recent, but less comprehensive
handbook is H. J. Bartsch, Handbook of Mathematical Formulas, translated from the 9th German edition by Herbert Liebscher, Academic
Press, New York, 1974. This work covers a great range of elementary
topics, from the simplest properties of complex numbers as far as
elementary differential equations, probability and statistics, and
mathematical programming.

It is often difficult to make a clear distinction between handbooks and tables as forms of literature. In general, when most of
the data is in tabular form the work is considered a book of tables.
Mathematical tables of logarithms, trigonometrical functions, square
roots, and so on, are designed to save the time and labor of those
engaged in computing work.

Mathematical tables date from the earliest historical records,
including the Almagest of Ptolemy, which contained tables of angle
chord values [3]. The advent of the computer has made possible the
construction of many special-purpose tables and has greatly accelerated the rate at which new tables are produced. The problem today
is to keep track of the many tables being published. The number of
compilations of mathematical tables designed for different purposes
and different classes of users is so extensive that no listing of
individual titles is attempted here. The remaining discussion focuses on sources for locating information about mathematical tables.

Two important guides that should be consulted by anyone seri-
ously trying to locate mathematical tables are:

Alan Fletcher, et al, Index of Mathematical Tables, 2nd edition,
Addison-Wesley, Reading, Mass., 1962. This is an index to well-
known tables of functions and to other less-known tables in books
and journals. Part 1 is an index according to functions; Part 2 is
an alphabetical bibliography by author; Part 3 lists known errors in
published tables.

Mathematics of Computation, a quarterly journal published by the American Mathematical Society is the second source of information about tables. This is a continuation of Mathematical Tables and Other Aids to Computation, a quarterly journal established in 1943 by the National Research Council of the National Academy of Sciences to serve as a clearinghouse for information concerning mathematical tables. The Society has published an Index to Mathematics of Computation (1972), which is a compilation by author and subject of all material which appeared in Mathematics of Computation and its predecessor, Mathematical Tables and Other Aids to Computation during the years 1943-1969 (twenty-three published volumes).

Information about currently published tables can be located in journals and abstracting publications. Entries for tables are found in the monthly issues of Mathematical Reviews in the "Statistics" and "Numerical Analysis" sections, and in the index volumes under the heading "Tables."

VI.   HISTORY AND BIOGRAPHY

The history of mathematics is largely an account of the works of individuals, making it difficult to separate history and biography. Histories usually provide information about mathematicians' writings and they may serve to some extent as bibliographies, as well. Poggendorff's Biographisch-literarisches Handwörterbuch der exakten Naturwissenschaften, discussed in Chapter 11, is a work that combines features of histories, biographies, bibliographies, and encyclopedias. It should be noted that in most libraries expository histories and biographies are shelved with the circulating collection, while chronologies and biographical dictionaries are shelved with the reference books.

Histories of mathematics vary in their treatment of the subject. Some are straightforward chronological presentations, some concentrate on a particular historical period, others present the subject in a cultural setting, some are source books, offering selections of original writings.

E. T. Bell, Development of Mathematics, 2nd edition, McGraw-Hill, New York, 1945, is a good one-volume historical survey that provides an orientation and comprehension of the subject as a whole.

The multi-volume World of Mathematics, edited by J. R. Newman, 4 volumes, Simon and Schuster, New York, 1956, is another example of a work that covers mathematics from the earliest times to the present.

Other comprehensive histories include: D. E. Smith, History of Mathematics, 2 volumes, Ginn, Boston, 1923-25, reprinted Dover, New York, 1958; D. J. Struik, A Concise History of Mathematics, 3rd revised edition, Dover, New York, 1967; and, Morris Kline, Mathematical Thought from Ancient to Modern Times, Oxford University Press, 1972.

A rather unique book that provides an excellent introduction to mathematical ideas is R. Courant and H. Robbins, What is Mathematics? An Elementary Approach to Ideas and Methods, Oxford University Press, New York, 1941.

Another book of this type is T. L. Saaty, The Spirit and Uses of the Mathematical Sciences, McGraw-Hill, New York, 1969.

Morris Kline has written several books in which he shows the relationship of mathematics to general cultural development. They are: Mathematics in Western Culture, Oxford University Press, New York, 1953; Mathematics and the Physical World, Crowell, New York, 1959; and, Mathematics: A Cultural Approach, Addison-Wesley, Reading, Mass., 1962.

Salomon Bochner's The Role of Mathematics in the Rise of Science, Princeton University Press, Princeton, N.J., 1966, traces the historical relationship of mathematics to scientific development.

The first few chapters of Mathematics: Its Content, Methods, and Meaning, edited by A. D. Aleksandrov, A. N. Kolmogorov, and M. A. Lavrent'ev, and translated by S. H. Gould and T. Bartha, M.I.T. Press, Cambridge, Mass., 1963, discuss the general cultural importance of mathematics and its historical development. Later chapters in this three-volume work discuss individual branches of contemporary mathematics.

A history in a different format is D. E. Smith, A Source Book
in Mathematics, McGraw-Hill, New York, 1929, reprinted, 2 volumes,
Dover, New York, 1959. This is a selection of 125 important writings
by eminent mathematicians, grouped under broad subject headings
(algebra, geometry, etc.).

Biographical information about mathematicians may also be found
in various forms and in different sources [1]. Famous mathematicians
may have a whole book devoted to them. Two examples are: S. R.
Ranganathan's biography of the Indian mathematician Ramanujan,
Ramanujan: the Man and the Mathematician, Asia Publishing House,
New York, 1967; and, Courant in Göttingen and New York, by Constance
Reid, Springer-Verlag, Berlin, 1976, the biography of Richard Courant.

Many more mathematicians are subjects of individual chapters
in works of collected biography. E. T. Bell, Men of Mathematics,
Penguin Press, New York, 1953, and Some Nineteenth Century British
Scientists, edited by R. Harré, Pergamon Press, Oxford, 1969, are
two examples. (The latter book includes mathematicians.)

For major figures there are sometimes other sources that are not
themselves primarily biographical, but which do provide biographical
details. Collected works of an individual, for example, often include information of an introductory or explanatory nature on the
author's life. John von Neumann: Collected Works, edited by A. H.
Taub, Pergamon Press, Oxford, 1961-63, is an example.

Festschriften, dedicatory volumes of papers by colleagues or
former students, usually include background information about the
individual; for example, Alex Heller and Myles Tierney, editors,
Algebra, Topology, and Category Theory: A Collection of Papers in
Honor of Samuel Eilenberg, Academic Press, New York, 1976.

A similar type of publication is the commemorative volume of
papers dedicated to the memory of an individual. These may appear
as a special issue of a journal, as in the case of John von Neumann,
1903-1957, which is Volume 64, Number 3, Part II, May 1958, of the
Bulletin of the American Mathematical Society.

Information about well-known living mathematicians may be found
in biographical dictionaries of the "who's who" type. Such works

may be international in scope, as the McGraw-Hill Modern Men of Science, 1966-68, or they may be compiled on a national basis, as American Men of Science, 1st-11th editions, 1906-68, Jacques Cattell
Press. The 12th edition of this dictionary, published by Bowker
in 1971, is titled American Men and Women of Science.

Most mathematicians belong to one or more of the national professional societies and they may be identified through society membership directories. The annual Combined Membership List of the
American Mathematican Society, The Mathematical Association of America, and the Society for Industrial and Applied Mathematics is an
example of such a directory. The main section of this directory is
an alphabetical listing of members, giving only factual data, such
as place of employment and mailing address. A geographic listing
arranges the names of members under three subdivisions: United
States, Canada, and other foreign countries. Examples of more specialized directories are:

Directory of Women Mathematicians, American Mathematical Society,
Providence, R.I., 1973. This directory of women Ph.D.'s and Ph.D.
candidates includes the name(s) by which the mathematician is known,
her address, present position, degrees, field(s) of interest, and
bibliography. There are yearly supplements to the directory.

Mathematical Sciences Administrative Directory, American Mathematical Society, Providence, R.I. This annual directory lists chair-
men of departments of mathematics in the United States, Canada,
Central America, and the Carribean; heads of nonacademic research
groups; heads and key personnel of government agencies; editors of
mathematics journals; officers and committee members in professional
mathematical societies.

World Directory of Historians of Mathematics, edited by K. O.
May and C. M. Gardner, University of Toronto, 1972. This is a list
of about 700 historians of mathematics, arranged in alphabetical order
and indexed by field of interest and by geographical location. Revision policies are not known at this time.

Reviews of new publications on mathematical history and biography appear in Mathematical Reviews under the heading "History and

Biography." Articles on historical research in mathematics, as well as book reviews, are included in <u>Historia Mathematica</u>, a quarterly journal, published since 1974 by the University of Toronto Press.

REFERENCES

1. Denis Grogan, <u>Science and Technology: An Introduction to the Literature</u>, 2nd edition, Linnet Books, Hamden, Conn., 1973.

2. A. J. Lohwater, "Mathematical Literature," <u>Library Trends</u>, 15, 852-867. (1967).

3. D. J. Struik, <u>A Concise History of Mathematics</u>, 3rd revised edition, Dover, New York, 1967.

Chapter 13

APPLIED MATHEMATICS

I. GENERAL

Mathematics done for its own sake is traditionally designated as
"pure mathematics," and mathematics concerned with problems arising
outside of mathematics itself is classified as "applied mathematics."
Such a division of mathematics into pure and applied is difficult to
maintain, however. The origin of many mathematical ideas important
in the purest branches of mathematics can be traced to applications
(e.g., Fourier series) and, on the other hand, mathematics created
for its own sake often turns out to be important for applications
(e.g., non-Euclidean geometries) [1].

Another common misconception is that there is a single field
called "applied mathematics" having the same unity of background and
historical tradition of mathematics as a whole. The current vari-
eties of application of mathematics are quite different from one
another in context and method and are not based upon any common
methodological core. Applied mathematics might be most accurately
described as a collection of separate fields, each of which starts
with the methods of one or another mathematical subfield and concerns
itself with a field of application external to mathematics proper [2].

The degree of mathematization, the sophistication of mathematical tools used, and the lasting intellectual value achieved by
the use of mathematics vary widely from field to field.

Computer science and statistics are fields that have a two-part
nature, only one part of which is mathematics. Neither could exist

without mathematics, but computer science would be unproductive
without the machine, and statistics requires the quantitative aspects of the scientific method. The range of applications of these
two fields is almost as wide as that of mathematics itself.

Physical mathematics (classical applied mathematics) and operations research differ from computer science and statistics in that
they are fields that have derived their nature from some area of
application. While the mathematical tools used in these fields were
initially chosen to meet the problems of an area of application,
the fields have since grown and developed enough to have a mathematical character of their own.

Alongside these four main applied mathematical sciences there
are still newer areas of application where no self-identifying
community of mathematical scientists yet exists. Fields such as
mathematical biology, mathematical psychology, and mathematical lin-
guistics, among others, deal with the mathematical aspects of rather
specific areas [1].

In view of this diversity, it is understandable that there is
no separate body of primary literature that can be labeled "applied
mathematical literature," and no unified system of secondary publications providing access to the published work in applied mathematics. Most of the individual applied mathematical sciences do
have their professional societies, their primary journals, and in
some cases, their abstracting journals. However, there is almost no
coordination among these sciences. Surveys of mathematicians in the
individual fields of applied mathematics show varying degrees of
satisfaction with the literature and with access to it. As might be
expected, people in the older, well-established fields are better
served than people in the newer fields of application [3].

The following sections of this chapter deal with the professional
societies of applied mathematicians, some of the literature that is
more general in nature (that is useful to several of the fields of
applied mathematics), and some of the problems that are common to
all of the fields. Chapters 14 and 15 deal with two specific fields:
statistics, and operations research. These two have been chosen to

illustrate the two different types of applied mathematical sciences
that were described above. In addition, statistics is an example
of one of the older applied mathematical sciences and operations research is one of the newer fields. The discussion of each field and
its literature is merely indicative, since each field warrants a
complete guide of its own.

II.   PROFESSIONAL SOCIETIES

The Society for Industrial and Applied Mathematics, usually referred to as SIAM, is the major society for those concerned with
mathematics and its applications. It was founded in 1952 to further
the applications of mathematics to problems in industry and science,
and to foster cooperation between mathematics and other sciences and
technology. It is probably the most comprehensive of all the professional societies of applied mathematicians, in terms of the variety
of interests of its members and the range of its activities. Like
the American Mathematical Society and The Mathematical Association
of America, SIAM conducts meetings at the national and regional levels,
sponsors a visiting lectureship program, and has an extensive publication program to provide for the exchange of ideas among all who
are interested in the applications of mathematics.

Some representative societies of a more specialized nature
than SIAM are:

The American Statistical Association, founded in 1839, which is
the oldest of the professional societies, including the American
Mathematical Society.

The Institute for Mathematical Statistics, founded in 1935,
which is an international organization devoted to the development,
dissemination, and application of mathematical statistics.

The Association for Computing Machinery, which was organized
in 1947 to advance the sciences and arts of information processing
and promote the exchange of information about them.

The Operations Research Society of America, which was organized
in 1952 to advance this field through the exchange of information,

establishment and maintenance of standards, and encouragement and
development of students.

The Institute of Management Sciences, which was founded in 1955,
is another international society whose objective is to identify,
extend, and unify scientific knowledge pertaining to management.

III. THE PRIMARY LITERATURE

The literature of the applied mathematical sciences, like that
of research mathematics, falls into two broad categories: the primary literature, and the secondary literature. The primary literature,
consisting of journals, proceedings of symposia, and monograph
series is discussed below. The secondary literature, abstracts and
other reference books, is the subject of the following section.

A. Journals

Currently, eight of the primary journals are published by the
Society for Industrial and Applied Mathematics. Since some of these
are general in scope and others are specialized, they provide good
examples of the kinds of journals available in the applied mathematical sciences and are described in some detail.

SIAM Journal on Applied Mathematics (1953-). This journal
covers three areas of applied mathematics: Part A. The applied
mathematics of the physical and engineering sciences; Part B. Discrete mathematics, probability, statistics, and operations research;
Part C. Biological and societal problems. There are two volumes
per year, four issues per volume.

SIAM Journal on Computing (1972-). This quarterly journal contains the latest and most up-to-date articles in mathematics applying
to computer science problems and other nonnumerical aspects of computing. Topics include automata theory, analysis of algorithms, computational complexity, computational aspects of combinatorics and
graph theory, the mathematical aspects of programming languages,

artificial intelligence, information retrieval, data structures, and
computer architecture.

SIAM Journal on Control and Optimization (1963-). This bimonthly journal contains research articles in the mathematical theory
of control and its applications, in associated areas of system
theory, in optimization, including continuous and discrete mathematical programming in the theory of games and theory of differential
games, and in those topics of mathematical analysis, applied probability, and stochastic processes that are directly related or
applicable to control theory or optimization.

SIAM Journal on Mathematical Analysis (1970-). This journal is
devoted to that part of analysis that bridges abstract pure mathematics and numerical, physical, and engineering applications. Topics
include approximation theory, asymptotic analysis, differential
equations, generalized functions, integral equations, integral transforms, and special functions. There are six issues per year.

SIAM Journal on Numerical Analysis (1964-). This journal focuses
on the field of development and analysis of numerical methods, including their convergence, stability and error analysis, along with
related results in functional analysis and approximation theory, as
well as the description of computational experiments and new types
of numerical applications. Six issues per year.

Theory of Probability and Its Applications (1956-) is a translation of the Russian journal Teoria Veroyatnostie Iee Primenenie.
A quarterly publication, it contains original papers and short communications on the theory of probability, general questions in mathematical statistics and applications of probability theory in social
sciences as well as engineering. It also contains papers on game
theory.

SIAM Review (1959-). This is also a quarterly journal, which
contains expository and survey papers in all areas of applied mathematics, as well as essays on topics of current interest to applied
mathematicians. Also included are sections of "Problems and Solu-
tions," current "Book Reviews," and "Classroom Notes in Applied
Mathematics."

SIAM News (1968-) is a bimonthly publication intended primarily to inform members and the scientific community of items of current interest, such as meetings, scholarships, and other news items.

Journals are also published by universities and commercial firms. The Quarterly of Applied Mathematics, published by Brown University, the Journal of Applied Probability, published by the University of Sheffield, and several of the journals published by Academic Press, Marcel Dekker, North-Holland Publishing Company, and Springer-Verlag have already been mentioned in Chapter 7. Additional titles may be located through the bibliographies and lists described in Chapter 8.

B. Books

Since 1967 the Society for Industrial and Applied Mathematics and the American Mathematical Society have co-sponsored an annual symposium in applied mathematics. Proceedings of these symposia are published under the title SIAM -AMS Proceedings. Each symposium is devoted to a single topic and each volume in this series contains the papers presented at the symposium.

Proceedings of Symposia in Applied Mathematics, 1947-1961. These volumes contain papers presented at symposia in applied mathematics held under the auspices of the American Mathematical Society and other interested organizations from 1947 to 1961. Although the latest catalog of American Mathematical Society publications states that this book series is superseded by the new SIAM-AMS Proceedings, there have been additions since Volume 14 (1961).

Lectures in Applied Mathematics (1957-). The volumes in this series contain lectures given at summer seminars sponsored by the American Mathematical Society on various topics in applied mathematics.

Lectures on Mathematics in the Life Sciences (1966-). The books published in this series contain papers presented at symposia held under the auspices of the American Mathematical Society and the American Association for the Advancement of Science.

CBMS Series. This is a series of monographs, each written by an individual author and published for the Conference Board of the Mathematical Sciences by the Society for Industrial and Applied Mathematics. The topics in this series are diverse, ranging from problems in the biological sciences to indexing theory.

Both the Carus Mathematical Monographs and The MAA Studies in Mathematics, which were described in Chapter 10, include topics in applied mathematics.

The MAA Cooperative Summer Seminar Notes (1971-). The volumes in this series contain lectures given at seminars sponsored by The Mathematical Association of America. The purpose of these seminars is to present contemporary mathematical topics to teachers of college mathematics with the ultimate goal of updating and upgrading college teaching. Each seminar and its corresponding volume is devoted to a single area, such as Mathematics in the Behavioral Sciences (1973).

All of the above series are published by societies. However, universities and commercial publishers, both in the United States and other countries, issue books in applied mathematics. Again, as is the case in research mathematics, these books are usually published as parts of series. Some representative series titles and their publishers are listed below.

Addison-Wesley Series in Applied Mathematics (Addison-Wesley Publishing Company).

Applications of Mathematics Series (Thomas Nelson and Sons).

Applied Mathematical Sciences (Springer-Verlag).

Applied Mathematics and Mechanics; An International Series of Monographs (Academic Press).

Cambridge Monographs on Mechanics and Applied Mathematics (Cambridge University Press).

Lecture Notes in Economics and Mathematical Systems: Operations Research, Computer Science, Social Science (Springer-Verlag).

McGraw-Hill Series in Modern Applied Mathematics (McGraw-Hill Book Company).

Mathematics and Its Applications (Gordon and Breach Science Publishers).

North-Holland Series in Applied Mathematics and Mechanics
(North-Holland Publishing Company).

Prentice-Hall Series in Applied Mathematics (Prentice-Hall).

These are only a few examples of the numerous book series that
are available. Not included in the listing are foreign language
publications and separately published monographs. Additional titles
may be located in Irregular Serials and Annuals, through the bibliographies described in Chapter 11, and in individual publishers'
catalogs.

IV. THE SECONDARY LITERATURE

Most of the reference tools described in Chapter 12 contain
material that would be useful to the applied mathematician. In
addition, there are publications of a more specialized nature that
must be consulted by workers in the applied mathematical sciences.
Some examples of these types of secondary publications are discussed
in the remaining sections of this chapter.

A. Abstracts

Access to the primary literature through abstracts and reviews
has traditionally been good in the areas of research mathematics,
especially through Mathematical Reviews and its German counterpart
Zentralblatt für Mathematik und Ihre Grenzgebiete. However, in
applied areas of the mathematical sciences, not only are papers
widely dispersed in the primary journals, abstracts of the papers
are also widely dispersed in the secondary journals with differing
points of view. If he is to locate papers in his area of interest,
the applied mathematician must consult several abstracting journals.
The abstracting journals most frequently used, and the coverage of
each, are described below.

Mathematical Reviews (1940-). This journal has never reviewed
papers that contain no new mathematics, no matter how important the

applicattons, and in 1966 imposed the stronger requirement that
papers in applied mathematics should be judged for inclusion strictly
on their mathematical merits. This resulted in a substantial drop
in the number of articles reviewed in certain applied areas, such as
quantum mechanics.

Applied Mechanics Reviews (1948-). This is also a review journal; that is, the abstracts tend to be critical. It covers mechanics
of fluids and solids; heat; selected papers on analog and digital
computation, including numerical analysis; selected papers on control theory and its applications; and papers dealing with various
combinations of these. The emphasis is on engineering-oriented
papers, versus mathematics-oriented papers.

Zentralblatt für Matematik und Ihre Grenzgebiete (1931-).
While this covers mostly pure mathematics, there is a substantial
coverage of applied mathematics.

Computer and Control Abstracts (1966-). (Alternative title:
IEE Abstracts - Series C: Computer and Control Abstracts). This
journal, which covers control theory, technology, and applications;
computer programming and applications; and computer systems and
equipment, is directed primarily to engineers. It is prepared by
staff abstractors, who write uncritical abstracts, rather than reviews.

Computing Reviews (1960-) is a review journal that covers
various areas of computer science, including numerical analysis and
non-numerical computer mathematics.

Other abstracting journals used by applied mathematicians (depending upon their field of application) are:

Physics Abstracts (1898-).

Referativnyǐ Žhurnal (Matematika, Mehanika, etc.) (1953-).

International Abstracts in Operations Research (1961-).

Quality Control and Applied Statistics (1956-).

Statistical Theory and Methods Abstracts (1959-).

These last three journals are discussed in Chapters 14 and 15.

B. Reference Books

Reference materials for the applied mathematical sciences fall
into the same categories that have been discussed in connection with
research mathematics: bibliographies, indexes, encyclopedias,
dictionaries, handbooks and tables, and so forth.

The Encyclopedia of Mathematics and Its Applications and the
Mathematical Handbook for Scientists and Engineers have already been
described in Chapter 12. Examples of some other general reference
tools are:

W. R. Freiberger, editor, International Dictionary of Applied
Mathematics, Van Nostrand, Princeton, N.J., 1960, reprinted Krieger,
New York, 1977. This work, which was compiled by experts in different fields, is in English, with indexes in French, German, Russian,
and Spanish.

U.S. National Bureau of Standards, Applied Mathematics Division,
Applied Mathematics Series, 1948-, Washington, D.C. This series
includes tables, manuals, and studies of interest to physicists,
engineers, chemists, biologists, computer scientists, and applied
mathematicians. It is a continuation of the Mathematical Tables
Project, which was financed by the Works Projects Administration of
the City of New York during the 1930s and 1940s.

Titles of more specialized reference materials may be found in
the guides [4] and [5] listed at the end of this chapter. The titles
discussed in Chapters 14 and 15 of this guide may be regarded as
representative of the types of reference materials available in all
areas of applied mathematics.

REFERENCES

1. Committee on Support of Research in the Mathematical Sciences,
    Report of the Committee, The Mathematical Sciences, National
    Academy of Sciences, Washington, D.C., 1968.

2. F. E. Browder, "Does Pure Mathematics Have a Relation to the Sciences?" American Scientist, 64, 542-549, (1976).

3. Committee on a National Information System in the Mathematical Sciences, Information Needs in the Mathematical Sciences, Conference Board of the Mathematical Sciences, Washington, D.C., 1972.

4. E. P. Sheehy, Guide to Reference Books, 9th edition, American Library Association, Chicago, 1976.

5. A. J. Walford, Guide to Reference Material, Volume 1, Science and Technology, 3rd edition, The Library Association, London, 1973.

Chapter 14

STATISTICS

Modern statistics, a science based on the mathematical theory of
probability, is one of the older and more firmly established of the
applied mathematical sciences. Statistics had its beginnings over
a century ago in the application of algebra and differential equations to problems arising in the analysis of experimental and obser-
vational data. Manipulation of large masses of data is still a
challenging problem, but the development of new statistical techniques and the invention of the computer have permitted a much
wider range of application, including nonexperimental observations
of behavioral science and medicine [1].

The diverse applications of statistics are reflected in a substantial and equally diverse amount of published literature, both
primary and secondary. This chapter describes some of the sources
of information in this field.

The two articles [2], [3], and the book [4], listed at the end
of this chapter provide good general introductions to modern statistical theory and the work that statisticians do. Articles in the
general encyclopedias may also be consulted for nontechnical discussions of the subject.

I. THE PRIMARY LITERATURE

A. Journals

A survey of the information needs and practices among a sample
of members of the American Statistical Association, the Biometric

Society, and the Institute of Mathematical Statistics identified the most widely used journals and described their readerships. The following statistical journals were considered to be the most important.

Journal of the American Statistical Association (1888-). This is the research journal of the Association. Published quarterly, it contains original papers and includes a review section. It is of most interest to statisticians who are concerned with methods and techniques, research, or design of experiments.

American Statistician (1947-). Also published by the American Statistical Association, this is a quarterly journal devoted to refereed articles and letters to the editor. Statisticians employed by government agencies and university students tend to be the greatest readers of this journal.

AMSTAT News (1974-). This publication is issued monthly (except July-August and September-October) and contains news items that formerly were contained in the American Statistician.

Annals of Mathematical Statistics was a quarterly journal published by the Institute of Mathematical Statistics (Baltimore, Md.) from 1930 to 1972. In 1972 it was superseded by two bimonthly publications: Annals of Statistics and Annals of Probability. These new journals, like their predecessor, are research journals.

The Royal Statistical Society (London) publishes two journals: the Journal (1952-) and Applied Statistics (1952-). Readers of the first journal are mainly research mathematicians and those engaged in the design of experiments. Readers of the second are those working in the field of econometrics.

Zeitschrift für Wahrscheinlichkeitstheorie und verwandte Gebiete (1962-) is published by Springer-Verlag. It is issued irregularly and contains papers written in German, English, or French that are research-oriented.

Technometrics (1959-) is a quarterly journal of statistics in the physical, chemical, and engineering sciences. Published jointly by the American Statistical Association and the American Society for Quality Control, it is geared to industrial statisticians.

Two journals devoted to the application of statistical methods
to the field of biological research (biometrics) are: Biometrics
(1945-) and Biometrika (1901-). Biometrics is the journal of the
Biometric Society (Raleigh, N.C.) and Biometrika is a British publication.

Additional titles may be found in Ulrich's International Periodicals Directory under the heading "Statistics" and under the
headings for the fields of application (e.g., Biometrics is listed
under the heading "Biology," with a cross reference from "Statistics").

B. Books

Current trends in statistics are reported in the Proceedings of
the Berkeley Symposia on Mathematical Statistics and Probability,
published by the University of California Press. The Berkeley Symposia have been held every five years since 1945 and the published
Proceedings of six symposia are currently available.

The American Mathematical Society publishes a series of translations from Russian for the Institute of Mathematical Statistics.
Entitled Selected Translations in Mathematical Statistics and Probability, it is similar in format to the American Mathematical Society Translations - Series 2 discussed in Chapter 10. The series
started in 1961 and presently consists of thirteen volumes. The
volumes are indexed in the Index to Translations Selected by the
American Mathematical Society, 2 volumes, 1966-1973.

The subject of statistics covers such a broad area of application that it would be difficult to list even a fraction of all the
books in the field. Instead, some sources for identifying monographs and texts are discussed in the following sections. In addition
to the secondary sources that are mentioned, many of the primary
journals contain announcements of new publications and book reviews.

II. THE SECONDARY LITERATURE

A. Abstracts

The bulk of the statistical literature is abstracted in Statis-tical Theory and Methods Abstracts (1959-). This is a quarterly journal, published for the International Statistical Institute by Oliver and Boyd, Edinburgh. International in scope, it attempts to give complete coverage of published papers in the field of statis-tical theory and contributions to statistical method. All articles in the following journals are abstracted: Annals of Mathematical Statistics, Biometrika, Journal of the Royal Statistical Society, Bulletin of Mathematical Statistics, and the Annals of the Insti-tute of Statistical Mathematics. Selective abstracting of other journals is done, including: Biometrics, Metrika, Metron, Inter-national Statistical Institute Review, Technometrics, and Sankhya (the Indian Journal of Statistics). The abstracts are all in English, regardless of the language of the original paper.

Other abstracting journals containing reviews of the statistical literature are: Mathematical Reviews (1940-) and Quality Control and Applied Statistics (1956-).

It should be pointed out that these abstracting journals are used chiefly by research oriented statisticians. The survey re-ferred to above found that working statisticians placed greatest value on their own files in seeking statistical information. Indeed, many were unfamiliar with the abstracting journals.

B. Indexes

The Current Index to Statistics has been published jointly by the American Statistical Association and the Institute for Mathe-matical Statistics since 1975. This is a computerized index to the statistical journal literature. Issued annually, it covers papers in core and related journals. It is arranged by authors, important words in titles, and by key words.

The American Statistics Index, published by Congressional
Information Services, Inc., is a comprehensive indexing and abstracting guide to the statistical publications of the United States
government. Since 1972 it has provided complete coverage of nearly
400 Federal statistical sources and publications from the early
1960s are included on a selective basis. Published monthly, each
issue consists of an index and abstract volume covering statistical
publications released by the Government during the previous month.
The indexes are cumulated quarterly and annually.

C. Bibliographies

Probably the most comprehensive bibliography of retrospective
literature is M. G. Kendall and A. G. Doig, A Bibliography of Statistical Literature, 3 volumes, Oliver and Boyd, Edinburgh, 1962-1968.
This bibliography covers the literature of probability and statistics
from the sixteenth century through the 1950s. Volume 1, published
in 1962, covers the period 1950-58; Volume 2, published in 1965,
covers the period 1940-49; and Volume 3, published in 1968, covers
the pre-1940 literature, with supplements for 1940-49 and 1950-58.
There are over 30,000 entries, arranged alphabetically by author.

International Statistical Institute, Bibliography of Basic
Texts and Monographs on Statistical Methods, 1945-1960, 2nd edition,
Oliver and Boyd, Edinburgh, 1963, is an example of a more selective
bibliography, as the title indicates. Bibliographic details for
each entry are followed by extracts from reviews which appeared in
the major statistical journals.

Another bibliography of the above type is O. K. Buros, editor,
Statistical Methodology Reviews 1941-50, Wiley, New York, 1951.
This work quotes reviews of approximately 350 books, written in
English and written or reviewed during the period. Earlier years
are covered by Research and Statistical Methodology Books and Re-
views, 1933-38 and The Second Yearbook of Research and Statistical
Methodology, Books and Reviews, published in 1938 and 1941, respec-
tively, by Gryphon, Highland Park, N.J.

Finally, for older materials, there is a publication that leads one to additional bibliographies of the literature. H. O. Lancaster, Bibliography of Statistical Bibliographies, Oliver and Boyd, Edinburgh, 1968. This covers personal bibliographies, subject bibliographies, and national bibliographies. Included are subject and author indexes.

Information about more recent bibliographies can be found in the standard reference tools discussed in Chapter 11, and, as previously noted, reviews of recent publications appear in many of the statistical journals.

D. Miscellaneous Reference Books

A useful dictionary in this field is M. G. Kendall and E. R. Buckland, A Dictionary of Statistical Terms, 3rd edition, published for the International Statistical Institute by Hafner, New York, 1971. The first two editions of this title included a glossary of Russian-English terms. This feature was omitted from the 3rd edition and published separately as:

Samuel Katz, Russian-English/English-Russian Glossary of Statistical Terms, Based on A Dictionary of Statistical Terms, by M. G. Kendall and W. R. Buckland, Oliver and Boyd, Edinburgh, 1971.

Statistical tables are an important reference tool of the statistician. Tables may be published in journals or they may be published separately. There are several useful guides and indexes to statistical tables that have been discussed in Chapters 12 and 13. An interesting compilation of tables is a series entitled Selected Tables in Mathematical Statistics, published for the Institute of Mathematical Statistics by the American Mathematical Society. Each volume in this series contains several sets of tables that are too long to be published in a journal, but too short for separate publication. The introductory material for each set discusses methods of computation, accuracy, methods of interpolation (when required), and applications, and gives numerical examples of the use of the tables. To date, three volumes in this series have been issued:

Volume 1 (1970, revised edition 1973), Volume 2 (1974), and Volume 3 (1975).

The American Statistical Association publishes an annual Directory of Statisticians and Others in Allied Professions. This directory lists members of the Association, the Biometric Society, and the Institute of Mathematical Statistics.

As was mentioned at the beginning of this chapter, the numerous applications of statistical techniques have resulted in a substantial and diverse amount of literature. The reader should be aware that library classification systems organize this literature dependent upon the way the material is treated. When the emphasis is upon the mathematical methods, per se, the literature is classified as mathematics. When the emphasis is on the field of application, the literature is classified with the field; for example, Biometrics is classified in the main class for Biology.

REFERENCES

1. Committee on Support of Research in the Mathematical Sciences, Report of the Committee, The Mathematical Sciences, National Academy of Sciences, Washington, D.C., 1968.

2. Jack Kiefer, "Statistical Inference," in The Mathematical Sciences: A Collection of Essays, M.I.T. Press, Cambridge, Mass., 1969, pp. 60-71.

3. William Krushal, "Statistics, Moliere, and Henry Adams," American Scientist, 55, 416-428 (1967).

4. J. M. Tanur, ed., Statistics: A Guide to the Unknown, Holden-Day, New York, 1972.

Chapter 15

OPERATIONS RESEARCH

With the aid of the computer, mathematical methods have penetrated
every form of human activity. Those that concern the science of
decision-making and its applications comprise operations research
(or operational research, as it is called in Great Britain). Essen-
tially, mathematical models are constructed of parts of or sometimes
entire systems and computers are used to determine optimal schedules
or plans. Operations research is distinguished from systems engineering is that it focuses on systems in which human behavior is important. Operations research makes use of the older disciplines, such
as mathematics, logic, and statistics, as well as more recent developments, such as communication theory, decision theory, cybernetics,
organization theory, the behavioral sciences, and general systems
theory.

Operations research had its beginnings during World War II in
the use of simple mathematical models and mathematical thinking to
study the conduct of military operations. After the war, the same
attitudes and approaches were applied to business and industrial
operations and management. Today, the emphasis of this rapidly
growing field is on solving problems of allocation (routing problems and scheduling problems are two types) and on a broad class of
operational applications of probability (inventory management and
improving of service of queues and waiting lines, for example).

Analysis of mathematical models of systems to be optimized has
stimulated such mathematical disciplines as linear, nonlinear, and
integral programming; network, graph, and matrix theory; queuing,

stochastic processes, reliability theory; dynamic programming and
control theory [1].

A good survey of the nature and scope of operations research and
how it draws on and contributes to various areas of mathematics may
be found in T. L. Saaty, "Operations Research," Science, 178, 1061-
1070 (1972).

An account of the early history of operations research is given
on pages 3-35 in J. F. McCloskey and F. N. Trefethen, editors, Operations Research for Management, 2 volumes, John Hopkins Press,
Baltimore, Md., 1954-1956.

I. THE PRIMARY LITERATURE

A. Journals

Many of the journals which deal with operations research are
published by professional societies. They report the results of research and application, review new literature, and give information
about meetings and other activities of interest to the membership.

The oldest journal in the field is Operational Research Quarterly
(1950-), published for the Operational Research Society (London) by
Pergamon Press. The United States counterpart is Operations Research,
which is the journal of the Operations Research Society of America.
The journal was founded in 1952 and is issued six times a year. It
is devoted primarily to original papers. The Society publishes information about national meetings and other news items in its semiannual Bulletin (1952-).

Mathematics of Operations Research (1976-) is a quarterly journal published by the Institute of Management Science, which is an
international society.

SIAM Journal on Control and Optimization (1953-). This journal
contains research articles in the mathematical theory of control and
its applications, in associated areas of system theory, in optimization, including continuous and discrete mathematical programming in
the theories of games and of differential games, and in those topics

of mathematical analysis, applied probability, and stochastic processes that are directly related or applicable to control theory or optimization.

Part B of the SIAM Journal on Applied Mathematics (discussed in Chapter 13) is devoted to discrete mathematics, probability, statistics, and also to operations research.

Applied Mathematics and Optimization, An International Journal (1974-) is an example of a journal in this field that is issued by a commercial publisher. Published by Springer-Verlag, it is a quarterly journal, with articles chiefly in English. Its primary aim, according to the publisher, is to publish papers treating applied (practical) problems of optimization without compromising mathematical precision. The problems may embrace a wide diversity of areas, and papers are particularly welcome that deal with modeling and identification of systems.

Additional journals may be found in Ulrich's International Periodicals Directory. However, there is no heading "Operations Research" in the current edition of Ulrich. Titles of journals in that field are listed under the headings "Mathematics" and "Computer Technology and Applications."

B. Books

Since 1957 an International Conference on Operational Research has been held every three years. The published Proceedings of these conferences are an important source for discovering current trends and developments in the field of operations research.

An annual review of progress in the field is also available: Progress in Operations Research, Wiley, New York, v.1-, 1961-.

A book series sponsored by the Operations Research Society of America is published under the title Publications in Operations Research.

Titles of some other series are:

Operations Research and Industrial Engineering (Academic Press).

Applications of Mathematics (Springer-Verlag).

Applied Mathematical Sciences (Springer-Verlag).

Econometrics and Operations Research (Springer-Verlag).

Reviews of new books may be found in many of the journals of operations research as well as in the abstracting journals mentioned in the next section.

II. THE SECONDARY LITERATURE

In 1959 the Societe Francaise de Recherche Operationnelle (France), the Operational Research Society (Great Britain), and the Operations Research Society of America (United States) founded the International Federation of Operational Research Societies. Today the Federation has grown to include over twenty national societies. Profiting from the experience of older sciences in disseminating information about their field, the Federation turned its immediate attention to methods of compiling and distributing international abstracts [2].

In 1961 the International Federation of Operational Research Societies established the International Abstracts in Operations Research. This publication provides world-wide coverage, with reviewers in the member societies of the Federation analyzing the literature of their respective countries. The abstracts are written in English, arranged in classified order, with subject and author indexes for each issue, and cumulative indexes. The journal is issued quarterly and is published for the Federation by the North-Holland Publishing Company of Amsterdam.

Other journals that review or abstract selected literature in this field are Mathematical Reviews, Zentralblatt für Mathematik und ihre Grenzgebiete, Computer and Control Abstracts, and Applied Mechanics Review. These journals have been described in earlier chapters of this guide.

The need for retrospective indexing of the literature (prior to the start of the International Abstracts in Operations Research) is met in large part by:

J. H. Batchelor, Operations Research: An Annotated Bibliography, St. Louis University Press, 1959.

Case Institute, Operations Research Group, A Comprehensive Bibliography on Operations Research, Wiley, New York, 1958.

Finally, the Operations Research Society of America publishes an annual Directory of its members.

REFERENCES

1. Committee on Support of Research in the Mathematical Sciences, Report of the Committee, The Mathematical Sciences, National Academy of Sciences, Washington, D.C., 1968.

2. J. E. Pemberton, How to Find Out in Mathematics, 2nd edition, Pergamon Press, Oxford, 1969.

INDEX

Abel, N. H., 18
Abstracting journals, 67-80
  applied mathematics, 118-119
  operations research, 134
  statistics, 126-127
Abstracting Services, 80
Abstracts, 62-65 (see also
    Abstracting journals)
Ahmes Papyrus (see Rhind
    Papyrus)
Aleksandrov, A. D., 30, 107
Alexandria, 10
Algebra, 17-18
  derivation of word, 11
Al-Khowârizmî, Mohammed ibn
    Musa, 11
Almagest, 10, 105
American Mathematical Monthly,
    54
American Mathematical Society,
    52
  abstracting journals, 67-76
  journals, 52-53
  monograph series, 82-83
  proceedings of symposia,
    83, 116
  review volumes, 85-86
  translation journals, 53
  translations series, 84-85
American Statistical Associa-
    tion, 113
  journals, 124
American Statistics Index, 127
AMS (see American Mathe-
    matical Society)
AMS(MOS) Subject Classifica-
    tion, 70, 71, 76, 93

Analysis, 14, 16, 18-19
Analytic geometry, 13, 19
Analytic mechanics, 16
Annotated Bibliography of
    Expository Writing in
    the Mathematical Sciences,
    3-4
Apollonius of Perga, 10
Applications of mathematics,
    26-27, 111-113
Applied Mechanics Reviews, 119
Arabic mathematics, 11-12
Archimedes, 10
Arithmetica of Diophantus, 10
Aslib Book List, 96

Babylonian mathematics, 8
Besterman, Theodore, 92
Bibliographic Index, 92
Bibliographies (see also
    Catalogs):
  books, 3, 91-95
  dictionaries, 104
  expository literature, 3-4
  handbooks and tables, 105-106
  journals, 57-61
  operations research, 135
  statistics, 127-128
Biographisch-literarischen
    Handwörterbuch der exakten
    Naturwissenschaften, 92, 106
Biography, 106, 108-109
Boetius, Anicius, 11
Bolton, H. C., 61
Bolyai, Janos, 19
Book reviews, 3, 95-96
  statistics, 127

137